Eye and Vision Research Developments

Pharmacological Treatment of Ocular Inflammatory Diseases

EYE AND VISION RESEARCH DEVELOPMENTS

Additional books in this series can be found on Nova's website under the Series tab.

Additional E-books in this series can be found on Nova's website under the E-books tab.

EYE AND VISION RESEARCH DEVELOPMENTS

PHARMACOLOGICAL TREATMENT OF OCULAR INFLAMMATORY DISEASES

TAIS GRATIERI
RENATA F. V. LOPEZ
ELISABET GONZALEZ-MIRA
MARIA A. EGEA
AND
MARISA L. GARCIA

Nova Biomedical
Nova Science Publishers, Inc.
New York

Copyright © 2010 by Nova Science Publishers, Inc.

All rights reserved. No part of this book may be reproduced, stored in a retrieval system or transmitted in any form or by any means: electronic, electrostatic, magnetic, tape, mechanical photocopying, recording or otherwise without the written permission of the Publisher.

For permission to use material from this book please contact us:
Telephone 631-231-7269; Fax 631-231-8175
Web Site: http://www.novapublishers.com

NOTICE TO THE READER

The Publisher has taken reasonable care in the preparation of this book, but makes no expressed or implied warranty of any kind and assumes no responsibility for any errors or omissions. No liability is assumed for incidental or consequential damages in connection with or arising out of information contained in this book. The Publisher shall not be liable for any special, consequential, or exemplary damages resulting, in whole or in part, from the readers' use of, or reliance upon, this material.

Independent verification should be sought for any data, advice or recommendations contained in this book. In addition, no responsibility is assumed by the publisher for any injury and/or damage to persons or property arising from any methods, products, instructions, ideas or otherwise contained in this publication.

This publication is designed to provide accurate and authoritative information with regard to the subject matter covered herein. It is sold with the clear understanding that the Publisher is not engaged in rendering legal or any other professional services. If legal or any other expert assistance is required, the services of a competent person should be sought. FROM A DECLARATION OF PARTICIPANTS JOINTLY ADOPTED BY A COMMITTEE OF THE AMERICAN BAR ASSOCIATION AND A COMMITTEE OF PUBLISHERS.

Library of Congress Cataloging-in-Publication Data
Pharmacological treatment of ocular inflammatory diseases / Tais Gratieri
 [et al.].
 p. ; cm.
 Includes bibliographical references and index.
 ISBN 978-1-61668-772-4 (softcover)
 1. Eye--Inflammation--Chemotherapy. I. Gratieri, Tais.
 [DNLM: 1. Eye Diseases--drug therapy. 2. Anti-Inflammatory
Agents--administration & dosage. 3. Anti-Inflammatory Agents--therapeutic
use. 4. Drug Delivery Systems. 5. Inflammation--drug therapy. WW 166
P5355 2010]
 RE96.P53 2010
 617.7'061--dc22- 2010025684

Published by Nova Science Publishers, Inc. ✦ *New York*

Contents

Preface		vii
Abbreviations		1
Chapter I	Introduction	3
Chapter II	Mechanism of Inflammation	7
Chapter III	Novel Lipid-based Drug Delivery Systems for Ocular Inflammatory Diseases	11
Chapter IV	Conclusions and Future Trends	31
References		33
Index		51

Preface

Novel strategies for ocular anti-inflammatory therapies are extensively being investigated to increase drug bioavailability and decrease adverse side effects of common treatments. Colloidal carriers based on lipid materials are becoming a suitable alternative due to several advantages, including solubilization of hydrophobic/lipophilic anti-inflammatory drugs followed by bioavailability enhancement, modification of pharmacokinetic parameters, and protection of sensitive drugs from physical, chemical or biological degradation. Furthermore, by modulating the surface properties e.g. with viscosity-enhancing agents or mucoadhesive polymers may also improve drug bioavailability since the carriers are maintained in the target area for longer time. Likewise, submicron meter particles allow efficient crossing of biological barriers protecting the eye and transport enabling efficient drug delivery to the target tissues and fluids of the anterior segment (cornea, conjunctiva, sclera, anterior uvea) or posterior segment (uveal region, vitreous fluid, choroids and retina). Lipid-based colloidal carriers broadly comprise micro and nanoemulsions, liposomes and lipid nanoparticles (Solid Lipid Nanoparticles and Nanostructured Lipid Carriers). These have been already tested for ocular delivery of several anti-inflammatory drugs including corticosteroids and non-steroidal anti-inflammatory drugs, revealing a great potential discussed herein. Novel drug delivery systems are being improved on a daily basis since they are considered a promising strategy to enhance the ocular bioavailability of topically administered drugs.

Abbreviations

COX	Cyclooxygenase
FDA	Food and Drug Administration
HLB	Hydrophilic-Lipophilic Balance
IOP	Intraocular Pressure
LUV	Large Unilamellar Vesicles
NLC	Nanostructured Lipid Carriers
NSAIDs	Non Steroidal Anti-Inflammatory Drugs
MLV	Multilamellar Vesicles
OLV	Oligolamellar Vesicles
pI	Isoelectric Point
o/w	Oil-in-water
SLN	Solid Lipid Nanoparticles
SUV	Small Unilamellar Vesicles.

Chapter I

Introduction

Inflammation is the manifestation of vascular and cellular response of the host tissue to injury. Injury to the tissue may be inflicted by physical or chemical agents, invasion of pathogens, ischemia, and excessive (hypersensitivity) or inappropriate (autoimmunity) operation of immune mechanisms [1]. Regardless whether an inflammatory insult is the inciting agent or a result of another pathogenic mechanism, ocular inflammation may occur on a large and diverse group of conditions that could compromise human health. The classic signs and symptoms of inflammation, including itching, pain, redness, heat and swelling, often result from ocular inflammatory conditions such as giant papillary conjunctivitis [2], seasonal (intermittent) allergic conjunctivitis [3,4], uveitis [5], dry eye [6], as well as inflammation following injury and/or surgery [7,8]. Such clinical conditions are associated with local changes in blood flow and the invasion of immune cells and inflammatory mediators. If left untreated, ocular inflammation may lead to temporary or permanent vision loss [9].

The pharmacological approach to manage inflammation involves administration of anti-inflammatory agents, i.e., corticosteroids, non-steroidal anti-inflammatory drugs (NSAIDs), and other pharmacologically active compounds.

Corticosteroids are normally used to treat severe inflammation of external eye tissues, which includes inflammation following injury and/or surgery [7,8], uveitis [10] and severe cases of keratoconjuntivitis [11]. These drugs are very potent and effective, however, can produce a plethora of adverse ocular and systemic events, such as induction or exacerbation of glaucoma [12], tear-film instability, epithelial toxicity, progression of cataracts and increased risk of opportunistic infections [13,14].

NSAIDs have proven to be relatively safe in the topical management of ocular inflammations and can be used in less severe cases including for the surgically induced miosis [15,16], management of post-operative inflammation [17], treatment of allergic conjunctivitis [18], prevention and treatment of cystoid macular oedema [19,20] for in the control of pain associated with corneal abrasions [21,22]. There are, however, some reported adverse side effects include impaired corneal sensation [23], persistent epithelial defects [24], superficial punctate keratitis [25].

Despite the several effective anti-inflammatory agents available nowadays, the main challenge remains in the treatment of most ocular diseases by means of achieving local therapeutic concentrations of the drug, limiting its adverse side effects.

The eye is characterized by physiological barriers that limit drug entrance from blood circulation to its inner structures. These are the blood aqueous and the blood-retinal barriers [26]. As a consequence, systemic or oral drug therapy requires large drug dosages to reach the site of action in proper amounts, which may cause significant systemic side-effects [27].

Intravitreal, periocular and subconjunctival injections could minimize systemic exposure of the drug, but the use of these systems is followed by a series of disadvantages. The intravitreally injected drug is rapidly eliminated by the eye's natural circulatory process and therefore frequent injections may be required. Likewise, large dosages are often needed, giving rise to toxicological problems. Besides, there are also relevant side effects, e.g. pain, discomfort, increased intra ocular pressure, intraocular bleeding, increased chances for infection, and the possibility of retinal detachment. The major complication for intravitreal injection is endophtalmitis which can result in severe vision loss [28-30]. In addition, ocular injections are not well accepted by patients.

Ocular implants could represent and alternative to repeated injections as they are able to sustain drug release [31,32], however the insertion of these devices is invasive and can be followed by serious complications, such as retinal detachment, local haemorrhage (in case of intravitreal implants), cataract and increased intra ocular pressure [33]. The need of challenging surgical techniques has hampered the further development of these systems for routine clinical use.

The topical administration would be the preferred route for management of ocular inflammations, especially for inflammations affecting the anterior chamber structures. Nevertheless, unfortunately in several cases, the topical treatment is not effective enough due to protection mechanisms of the human eye, as lachrymal secretion and blinking reflex, which cause rapid drainage of the formulation [34]. Only 5% of the applied drug in conventional eye drops penetrates the cornea and reaches the intraocular tissues with the rest of the dose undergoing transconjunctival absorption or drainage via the nasolachrymal duct before transnasal absorption. This results in loss of drug into the systemic circulation and provides also undesirable systemic side effects [35]. Besides that, tight junctions between cells of the corneal epithelium provide an effective barrier against the penetration of most compounds. The short precorneal residence time allied with cornea impermeability results in low bioavailability and frequent dosing is usually needed to compensate the rapid precorneal drug loss.

Regardless the administration route many anti-inflammatory drugs do not possess the required physicochemical properties to be absorbed, and reach or enter target tissues. Most of the NSAIDs are weakly acidic drugs, which ionize at the pH of the lachrymal fluid and therefore have limited permeability through the anionic cornea which has an isoelectric point (pI) of 3.2. Reducing the pH of the formulation increases the unionized fraction of the drug which enhances permeation. However, being acidic, NSAIDs are inherently irritant [36], and reducing the pH of formulation further increases their irritation potential, and decrease their aqueous solubility [1]. Also the ocular administration of corticosteroids often results in insufficient drug concentration due to poor absorption, rapid metabolism and elimination, being the permeation through the cornea dependent on the drug partition coefficient and molecular-weight [37].

A promising strategy to overcome these problems involves the development of suitable drug carriers systems. The in vivo fate of the drug is no longer mainly dependent on by the properties of the drug, but on the carrier itself, which should allow a controlled and localized release of the active drug according to the specific needs of the therapy, whilst maintaining the simplicity and convenience of the dosage form. Various approaches, e.g. viscosity enhancement [38], use of mucoadhesive [39] or particulate [40] drug delivery systems, vesicular systems [41] and prodrugs [42,43], are being explored. Among these, lipid based drug delivery systems has received much attention due to the many advantages they can provide. Major advantages include: (i) prolonged and controlled action at the corneal surface; (ii) controlled ocular delivery by preventing the metabolism of the drug from the enzymes present at the tear/corneal epithelial surface; (iii) good biocompatibility due to the use of physiological and biodegradable lipids of low systemic toxicity; and (iv) for some formulations, the possibility of production on large industrial scale.

Considering this, the aim of the present chapter is to discuss the anti-inflammatory therapies available and describe the novel lipid based drug delivery systems such as microemulsions, nanoemulsions, liposomes and lipid nanoparticles, used for controlled release and drug targeting.

Chapter II

Mechanism of Inflammation

Intraocular inflammation is a clinical ocular disorder developed after eye injury by several biological, chemical or physical agents. Ischemia, hypersensitivity or autoimmunity causes may be involved. The ocular inflammation is involved in many pathological clinical conditions and other diseases affecting the anterior or posterior segment of the eye. Ocular structures, such as eyelids, conjunctiva, cornea (keratitis), and anterior or middle uvea (iritis, ciclitis) can be affected. The inflammatory response consists of miosis, lachrymation, conjunctival hyperaemia and breakdown of the blood aqueous barrier followed by protein leakage into the aqueous humour [44]. Inflammation in the back of the eye can be choroiditis (if affecting the uvea posterior), or retinitis (if affecting the retina). Vasculitis may also occur if the retinal vessels are inflamed.

Postoperative inflammation is also common after tissue injury upon surgical incision in cataract or other ophthalmic surgery. This incision triggers the inflammatory cascade, which begins with activation of phospholipase A2. The degree of postoperative inflammation following cataract surgery is linked to several surgery-dependent factors (e.g. surgical technique, intraocular lens type) and patient-dependent factors (e.g. history of inflammatory disease and degree of iris pigmentation). Extracapsular cataract extraction with posterior chamber lens implantation has become a safe procedure. However, during the surgery or postoperative period, specific complications related to prostaglandins and other inflammatory mediators may occure.g. acute pain and discomfort, intraoperative miosis, decrease in visual acuity, posterior capsule fibrosis, keratopathy, fibrin reaction, chronic uveitis, raised intraocular pressure (IOP), synechiae or secondary membrane.

Inflammation is regulated by a complicated mechanism responsible for the rupture of the blood ocular barrier and the attraction of leukocytes towards the eye. This cellular trafficking is regulated by the release of inflammatory mediators and cytokines. As a result of the injury, the phospholipids existing in the cell membranes are broken down into arachidonic acid, which is then converted to prostaglandins by cyclooxygenase (COX) or converted to hydroxy acids and leukotrienes by 5-lipoxygenase. The produced arachidonic acid enters either the COX or lipoxygenase pathway. Activation of COX pathway results in formation of prostaglandins and thromboxanes, whereas the lipoxygenase pathway yields eicosanoids (hydroxyeicosatetraenoic acid and leukotrienes).

Prostaglandins play an important role in the initiation and maintenance of ocular inflammation, being the mediators within the cellular and humoral inflammation cascade, including allergic reactions and pain response. These inflammatory mediators, which show chemokinetic activity, are present in the most tissues of the eye, being the conjunctiva and the anterior uvea the ocular structures which exhibit the most ability of synthesize prostaglandins. In the sclera, cornea, lens, choroid and retina this ability is weaker. Ocular prostaglandins released in the inflammatory process may act at different levels: prostaglandins E1 and E2 increase the IOP by local vasodilatation and increased permeability of blood aqueous barrier but prostaglandin F2 lowers the IOP by increasing uveoscleral outflow. At iris level, they act on the smooth muscle to cause miosis [1]. On the other hand, PGs cause vasodilatation and increase the vascular permeability resulting in increased aqueous humour protein concentration [45].

Corneal nociceptor terminals are excited by exogenous noxious stimuli and also by inflammatory substances released by damaged cells (arachidonic acid, metabolites, neuropeptides, biogenic amines, and kinins). Inflammatory mediators, such prostaglandins may directly activate corneal nociceptive terminals or induce sensitization, and contribute to the sustained ocular pain that follow corneal damage [45]. Accordingly, anti-inflammatory agents are also prescribed as analgesic drugs to reduce postoperative pain after photorefractive surgery [46]. Moreover, the anti-inflammatory agents elicit their action at different levels of arachidonic acid cascade. Corticosteroids act by blocking the enzyme phospholipase A2 to inhibit arachidonic acid production, thereby preventing the synthesis of all the PGs, thromboxanes and eicosanoids. On the other hand, NSAIDs exert their anti-inflammatory action by inhibiting the enzymes COX-1 and COX-2.

Ocular inflammation of the anterior segment of the eye is generally managed by topical administration of drugs, which are also approved by FDA for postoperative use. To provide a better control of the inflammatory condition and to prevent undesired side effects (as miosis) during the surgical procedure itself, their clinical use may be both in pre and post operative approaches [36]. Anti-inflammatory agents have also been used in preoperative prophylaxis [47].

2.1. Corticosteroids

Corticosteroids are anti-inflammatory agents since they inhibit phospholipase A2 and subsequently inhibit both the COX and lipoxigenase pathways [48]. Steroids interact with specific DNA sequences of cellular nucleus, changing the production of inhibitory proteins, and inhibiting the production of additional inflammatory mediators. Steroids also reduce the macrophages and neutrophils migration to the inflamed area, decreasing vascular permeability and suppressing the action of various lymphokines [49]. They also inhibit oedema, capillary dilatation, cellular infiltration, fibroblastic proliferation and deposition of collagen. Cortocisteroids are therefore the most current option to treat the inflammation associated with cataract surgery. They are recommended for preventing and/or treating postoperative ocular inflammation.

Since the corneal epithelium is lipophilic and the stroma is hydrophilic, molecules comprising both lipophilic and hydrophilic moieties (e.g., acetates or alcohols) penetrate the cornea to a greater degree rather than totally hydrophilic compounds (e.g., phosphate solutions) [50]. Prednisolone (acetate, phosphate) is most common topical corticosteroid, but others (e.g. fluocinolone acetonide [51], dexamethasone [52], fluorometholone [53], medrysone [54], and rimexolone [13]), have been used to treat inflammation, reducing adverse effects after surgical procedures [55]. These steroids can be used in ocular disorders that involve some inflammatory surface reaction such as dry eye, based on immune response and induced by many cytokines [6,56].

Corticosteroids offer the most potent efficacy in treating inflammation; however, they can induce significant side effects. Their short-term use helps minimizing the risk while still reaping significant benefits. Some of the most serious side effects of prolonged topical steroid application include ocular hypertension, glaucoma, cataracts, mydriasis, ptosis, inhibition of corneal epithelial or stoma healing, punctate staining, corneal-sclera melting, damage to the optic nerve, and defects in visual acuity and visual fields [57].

2.2. Non-steroidal Anti-inflamatory Drugs (NSAIDs)

NSAIDs comprise several chemically heterogeneous compounds which inhibit prostaglandins and thromboxane formation from arachidonic acid through the inhibition of the enzymes COX-1 and COX-2. Endogenous prostaglandins increase the permeability of the blood ocular barriers affect intraocular pressure and produce miosis and conjunctival hyperemia.

NSAIDs may be an effective alternative to costicosteroids in the topical management of ocular inflammations, being nowadays used in postoperative inflammation [58,59], inhibition of intra-operative miosis [60], treatment of seasonal allergic conjunctivitis [61], prevention and treatment of cystoid macular oedema [62], in pain relief [63], to decrease bacterial colonization of contact lenses and prevent bacterial adhesion to corneal cells [64].

NSAIDs for ocular inflammation include water soluble indoleacetic acid, aryl acetic acid, aryl propionic acid and enolic acid derivatives. The majority of the molecules are weak acids, which ionize at the pH of the lachrymal fluid and thus have limited permeability through the anionic cornea. Reducing the pH of the formulation increases the non-ionized fraction, enhancing drug permeation but increases the risk of irritation. There are some NSAIDs approved by the FDA for the treatment of post-operative inflammation after cataract surgery (kerotolac, flurbiprofen, bromfenac, diclofenac and nepafenac (a prodrug that is converted in its active form amfenac by intraocular enzymatic hydrolysis) [65]. These drugs are also prescribed for postoperative pain relief after photo-refractive surgery, being the analgesic action partially attributed to reduction of prostaglandins production [45].

The main advantage of using topical NSAIDs is the avoidance of undesirable effects of steroids, in particular for patients susceptible to corticosteroid-responsive IOP elevations, or with recurrent herpes simplex infection or delayed wound healing. Although NSAIDs are widely recognized as providing most of the clinical benefits of steroids while avoiding major adverse side effects, they also have limitations. These include ocular irritation and discomfort following application, conjunctival injection, mild punctate keratopathy, mydriasis, allergic and hypersensitivity reactions [36].

Chapter III

Novel Lipid-based Drug Delivery Systems for Ocular Inflammatory Diseases

Enhanced ocular retention of oily vehicles has been reported for more than 30 years [66], being attributed to their interaction with the superficial oily layer of the tear film. As a consequence, initial attempts to overcome the poor bioavailability of topically instilled drugs typically involved the use of ointments and emulsions. Ointments and emulsions ensured superior drug bioavailability by increasing the contact time with the eye, minimizing the dilution by tears, and resisting nasolachrymal drainage [67]. However, these vehicles have the major disadvantage of being uncomfortable and providing blurred vision. They still can be found for a series of anti-inflammatory drugs but are mainly used for either administration overnight or for treatment on the outside and edges of the eyelids [68]. Since only a limited percentage of the administered drug reaches the target tissue, patient compliance is an important aspect to consider when developing an ophthalmic delivery system. As such, attention should be paid to the facility of administration and to the sensorial feeling after the administration, since discomfort after administration (e.g., burning sensation), could induce tear production, followed by drug dilution and drainage through nasolachrymal duct.

Other important aspect to consider is the retention time, drug loading capacity and drug protection from metabolic degradation. In fact, if the system is able to prolong the retention, while loading a sufficient amount of drug in a protected manner, the interval between administrations can be lengthened. For instance, in case of intravitreal injections, the reduction on the number of injections would also reduce the potential side-effects. Apart from these, all the factors that would influence the overall costs should also be considered, as possibility of scaling up of production, sterilizing and the physical and chemical storage stability of the product.

The novel lipid based delivery systems as microemulsions, nanoemulsions, liposomes, SLN and NLC are now being used in attempt to attend all these requirements. The advantages and limitations of each system, and their suitability for the treatment of ocular inflammatory diseases will be reviewed in the following sections.

3.1. Microemulsions

The concept of microemulsions was first described in 1943 by Hoar and Schulman who generated a clear single-phase solution by titrating a milky emulsion with hexanol [69]. Microemulsions are defined as a system composed of water, oil and an amphiphile molecule which is a single optically isotropic and thermodynamically stable liquid solution' [70].

They are formulated using high concentration of surfactants, to decrease the interfacial tension at the oil/water interface and usually also co-surfactants (e.g., alcohol, amides and sulphoxides), to keep the interfacial layer highly flexible and fluid. In this way, a clear or translucent system composed of very small droplets (~100 nm) of oil or water is obtained stabilized by an interfacial film (Figure 1). These systems are called "Ternary systems" if only surfactants are used, or "Pseudoternary systems" if surfactants and co-surfactants are used together taken as a single-phase [71]. The ionic surfactants are generally too toxic to be used at high concentrations in ocular preparations, therefore, non-ionic surfactants are preferred [72].

The intrinsic properties of microemulsions confer special advantages to its use as ophthalmic drug delivery systems. Firstly, as they are thermodynamicaly stable they can be easily prepared and easily sterilized. Secondly, they have a high solubilizing capacity [74]. Besides, the surfactants and co-surfactants can act as penetration enhancers to facilitate corneal penetration of drugs [70] what could be further potentialized by the good spreadability on the cornea and mixing with the precorneal film constituents, due to the low surface tension of microemulsions [75]. Most likely due to their oily nature, microemulsions have demonstrated prolonged retention in comparison with aqueous solutions [72], which could reduce frequency of

Nanoemulsion strongly improved the solubilization of indomethacin and sodium diclofenac [74]. However, one limitation of this system is the low possibility of controlling drug release due to the small size and the liquid state of the carrier [88].

The earlier attempts on applying nanoemulsions for ocular delivery of anti-inflammatory drugs were made using anionic formulations. The in vitro corneal penetration of indomethacin from anionic nanoemulsions composed of lecithin and Miglyol 840 oil has demonstrated to be more than 3-fold that of the commercial eye drops. In addition, this property was comparable between nanoparticles and nanocapsules made of poly-ε-caprolactone with similar particle size of the nanoemulsions (200-250 nm), which therefore excludes the influence of the inner structure or chemical composition of the colloidal systems on the corneal penetration of indomethacin [89]. As further animal studies showed that microparticles containing indomethacin hardly increased drug penetration, the authors concluded that the main factor responsible for the favourable corneal transport of indomethacin was the colloidal nature of these carriers rather than their inner structure or composition [90]. Similar results (i.e., a 3.8-fold indomethacin permeation increase in comparison to commercially available eye drops), were obtained using an anionic nanoemulsion composed of 20% oily phase, 0.2-1% phospholipids and 0.2-0.5% amphoteric agents. In this formulation, the pH value was adjusted to 3.8 in order to maintain the indomethacin in the oily phase, preventing its ionization (pKa = 4.5). In this case, the increase of indomethacin permeation can be also attributed to the low ionization of the drug at emulsion pH. This resulted in a higher partition to the lipophilic epithelium layer compared to the indomethacin aqueous solution that was adjusted to pH 6.8, where the drug is markedly ionized [79]. On the other hand, in vivo the lower pH of the external phase would result in pH-induced lachrymation and loss of drug from the conjunctival sac resulting in reduced bioavailability and return of the pH of lachrymal fluid back to physiological range would reduce ocular penetration of drug due to ionization [1].

Another preparation made of 4.0% (w/v) of polysorbate 80, a non ionic surfactant and 5.0% (w/v) was capable of a 5.7-fold increase in the ocular penetration of difluprednate, a synthetic glucorticoid, in comparison with the drug ophthalmic suspension [81].

The scientific literature reports the occurrence of electrostatic interactions between the cationic emulsified droplets and anionic cellular moieties of the ocular surface [91]. In addition to the intrinsic corneal cells membrane negative charge, a layer of the glycoprotein mucin (a mixture of neutral and acidic mucopolysaccharides) secreted by goblet cells at the conjunctival surface is adjacent to the corneal epithelium, what makes it negatively charged with an isoelectric point of 3.2 [92]. The hypothesis of electrostatic interaction between positively charged emulsions and the cornea surface has been supported by various studies demonstrating that positive charge may prolong the residence time of the drop on the epithelial layer of the cornea and thus enable better drug penetration through the cornea to the internal tissues of the eye. Animal studies have demonstrated that the contact angle of one droplet of the different dosage forms on the cornea is found to be 70° for saline, 38° for an anionic nanoemulsion and 21.2° for a cationic nanoemulsion. The values of the spreading coefficient were found to be -47, -8.6 and -2.4 mN/m, respectively. It can be clearly deduced that both nanoemulsions had better wettability properties on the cornea compared to saline, being the positively charged superior than the negatively one. In these systems, it is likely that the drug is not released from the oil droplet in a hydrophilic tear compartment but rather partitioned directly from the oil droplets to the cell membranes on the corneal epithelium [82,83].

A positively charged nanoemulsion containing piroxicam demonstrated to be effective in lowering the ulcerative cornea score, following alkali burn of rabbit corneas [93]. Likewise, the comparison of indomethacin corneal penetration from the positively charged nanoemulsion Indocollyre® (a marketed hydro-PEG ocular solution) and a negatively charged nanoemulsion revealed better performance of the positively charged formulation, as its spreading coefficient on cornea was four times higher than that of the negatively charged emulsion [94].

A study evaluating the residence time of indomethacin after instillation in rabbit eyes compared the performance of chitosan-coated emulsion with a non-coated emulsion. The coated emulsion had a mean particle size of 117.6 nm and positively charged (zeta potential of 27.7 mV), whereas the non-coated emulsion was of smaller mean size (94.8 nm) and slightly negatively charged (-6.2 mV). Coated emulsion provided mean concentrations 3.6-fold and 3.8-fold higher than the non-coated at 0.5 hr and 0.75 hr after instillation, respectively. The drug levels in cornea, conjunctiva, and aqueous humour 1 hr after instillation were also higher than those obtained after administration of the non-coated emulsion. After in vitro mucoadhesive tests the authors concluded that the residence time of the emulsion in tear fluid was attributed to the mucoadhesive properties of chitosan [95].

Chitosan nanoparticles were also compared to positively charged chitosan nanoemulsions for ocular indomethacin delivery. In vivo studies and histopathological examination revealed that rabbit eyes treated with nanoemulsion showed clearer healing of corneal chemical ulcer with moderate effective inhibition of polymorph nuclear leuckocytic infiltration, compared with nanoparticles preparation. Using nanoemulsions, therapeutic concentrations of indomethacin were achieved in the cornea and were significantly higher than those obtained following instillation of indomethacin solution [80].

3.3. Liposomes

Liposomes are phospholipid vesicles formed by one or several lipid bilayers. In each bilayer the nonpolar fatty acid tails are placed in the interior, whereas the polar heads are turned outside containing an aqueous phase both inside and between the bilayers (Figure 2). As such, appearance and permeability of phospholipid layers are similar to those of biological membranes [96].

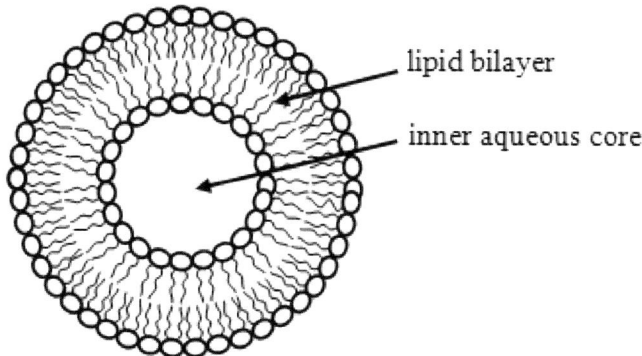

Figure 2. Schematic representation of a liposome (adapted from Bonacucina et al. [73]).

Liposomes were first introduced by Bangham et al. in the 1960s [97]. Studying the behaviour of lecithin and other phospholipids, the authors found that they are able to form spheres in dilute aqueous solutions. They defined liposomes as the smallest artificial vesicles of spherical shape that can be produced from natural nontoxic phospholipids and cholesterol. Bangham et al. were the first to study the physicochemical properties of these colloidal mesophases, such as their particle size distribution, osmotic behaviour, and surface charge [73].

Liposomes are formed spontaneously when phospholipids are hydrated in aqueous media. The mean size can vary from 20 to above 1000 nm, depending on composition, preparation method and number of bilayers. According to the number of bilayers they are distinguished as small unilamellar vesicles (SUV) (25-50 nm), large unilamellar vesicles (LUV) (100-200 nm), oligolamellar vesicles (OLV) and multilamellar vesicles (MLV) (1-2 µm) (Figure 3).

Since liposomes are composed of similar substances as cell membranes it is expected that they are biocompatible and biodegradable preparations. Most liposomes are prepared by using lecithin of egg or vegetable (soy bean) origin. Furthermore, a common component used in liposomes preparations is cholesterol, which is applied to improve characteristics (e.g., fluidity), to reduce the permeability of water-soluble molecules through the membrane in order to control release and to increase the stability of the bilayer membrane in the presence of biological fluids (e.g., blood/plasma) [96].

Several factors can influence the fate of drugs in liposome ocular delivery, namely: (i) the chemical composition of the liposomal product (lipids, surfactants, and other molecules); (ii) mean size of the vesicle; (iii) surface charge; (iv) drug-liposome interaction; and (v) the production process. Because of their amphiphilic character, liposomes are able to entrap both hydrophilic and hydrophobic compounds in the aqueous compartments or within the lipid bilayers, respectively [98,99]. Nevertheless, the encapsulation efficiency is generally higher for lipophilic rather than for hydrophilic molecules [96]. Liposomes can provide controlled release of incorporated drugs since the spherical lipid shield formed by bilayer membranes provides a permeability barrier to drug release. In this way, the drug is protected from degradation and clearance, and toxicity resultant from high peak concentrations is avoided. This property can be especially useful for posterior segment applications [100].

The increased residence time of drugs, and maintenance of their therapeutic concentrations for longer time intervals, could reduce the number of subconjunctival and intravitreal injections usually required in some treatments [101-104], while allowing higher doses without toxicity from initial concentration [84]. In this way, liposomes can minimize some of the adverse side effects encountered by these administration routes, increasing therapeutic effectiveness when no other options are available. Intravitreal injection of liposomes containing antibiotics [105-109], antifungal agents [110], antiviral agents [111-114], immunosuppressives [115], oligonucleotides [116,117] and other pharmaceutically active compounds, have been evaluated, but unfortunately not the same effort has been found for liposomes containing anti-inflammatory drugs.

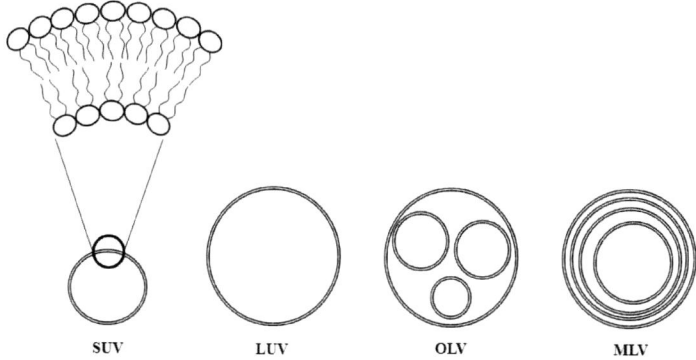

Figure 3. Schematic representations of different types of liposomes, depending on the shape, size and number of bilayers: small unilamellar vesicles (SUV), large unilamellar vesicles (LUV), oligolamellar vesicles (OLV) and multilamellar vesicles (MLV) (adapted from Bonacucina et al. [73]).

In a recent study, the potential of target-delivering corticosteroids to inflamed eye was suggested using intravenous injection of a sugar chain modified liposome containing dexamethasone in experimental autoimmune uveoretinitis mice. The authors verified that while intravenous administration of the drug alone provided a wide distribution in all tissues (eye, brain, heart, lung, liver, kidney, spleen and intestine), the intravenous injection of the modified liposome was almost concentrated to the eye, even though drug concentration in liposome formulation was the double of the formulation containing drug alone [118].

For the anterior segment, topical ocular administration of drugs incorporated into liposomes also showed several advantages, as increased corneal adherence [119], therefore, prolonged residence time [120], and improved bioavailability. Moreover, the controlled release allows prolonged effect [121], therefore time between administrations can be lengthened. In addition, liposomes offer the convenience of an ophthalmic drop and confinement of the action at the site of administration [122].

The use of liposomes for the topical delivery of anti-inflammatory agents has been investigated since 1983, when Singh et al. encapsulated triamcinolone acetonide in large MLV composed of phosphatidylcholine and cholesterol. These liposomes produced higher drug levels in ocular tissues of rabbits in compared to a standard suspension of the steroid. A more than twofold drug concentration was obtained and maintained in the aqueous humour for up to 5 h when using the liposome formulation [123].

Probably due to fact that topical steroids poorly penetrate the cornea with intact epithelium, first studies on liposomes for topical delivery focused on the application of these compounds. The efficacy of liposomes containing dexamethasone and its ester was investigated and compared to similar aqueous suspensions of each drug. The authors verified that the liposome containing dexamethasone valerate provided the highest ocular drug levels. In the case of dexamethasone or dexamethasone palmitate, the liposomes provided a lower drug level in comparison with the suspension. In this study, a high esterase activity was observed in the corneal homogenate supernatant, and most of the steroid taken up after instillation of dexamethasone valerate was metabolized to free alcohol [124].

Radiolabeled cationic liposomes were used for dexamethasone delivery [125,126]. This corticosteroid achieved high concentrations in the rabbit cornea, aqueous humour, iris, ciliary body, and sclera for up to 4 hours. Drug levels remained high for 6 hr in the conjunctiva. The authors anticipated that the saturated phospholipids in liposomes may protect them from tear lysozyme and esterase activity, resulting in greater bioavailability.

Diclofenac sodium was also loaded in cationic liposomes prepared by reverse phase evaporation [127]. They were shown to provide a 211% increase in aqueous humour concentration of drug compared with conventional aqueous eye drop formulation. In another study, notable advantages for ocular drug delivery were achieved by coating liposomes containing sodium diclofenac with chitosan. The authors reported that this procedure conferred a positive charge and slightly increased liposomes particle size, while the drug encapsulation was not affected. The coating procedure also prolonged in vitro drug release profile, improved physicochemical stability, and increased bioadhesion. This resulted in a prolonged retention compared to the non-coated liposome or drug solution, and displayed a potential penetration enhancing effect for transcorneal delivery. In the ocular tolerance study, no irritation or toxicity was caused by continual administration of low molecular weight chitosan-coated liposome in a total period of 7 days [128].

Significant progress has been made in demonstrating the advantages of liposome-mediated drug delivery in ophthalmology. In some cases, liposomes have shown to improve efficacy, reduce toxicity, prolong activity and provide site specific delivery. Despite those reasons, which make liposomes a potentially useful system for ocular delivery, until nowadays there were very few attempts on applying them for the treatment of ocular inflammatory diseases. Problems normally encountered were short shelf life, limited drug loading capacity, use of aggressive conditions for preparation, and problems in sterilization [129]. Temperatures required for autoclaving can cause irreversible damage to vesicles while filtration reduces the vesicle to an average of 200 nm limiting its use to small vesicles.

Several anti-inflammatory drugs have been successfully incorporated in liposomes, particularly in the last decade, e.g., aceclofenac [130], piroxicam [131,132], salicylic acid [133], indomethacin [134], ibuprofen [135], flurbiprofen [136], ketoprofen [137], dexamethasone [138], and betamethasone [139,140]. In some cases high encapsulation efficiency and long-term stability during storage were achieved [136,141]. Therefore, as these main formulation problems are being solved it is expected that in the next years studies applying liposomes containing anti-inflammatory agents for ocular drug delivery will increase massively and therefore more therapeutic options will become available.

3.4. Lipid Nanoparticles (SLN and NLC)

Solid lipid nanoparticles (SLN) are the first generation of nanoparticles composed of lipids that are solid at room and body temperatures, stabilized with an emulsifying layer in an aqueous dispersion, i.e.,, they resemble the nanoemulsions by replacing the inner liquid lipid with a solid lipid. They were developed in the beginning of the 90s. The main advantage of SLN over nanoemulsions is the possibility of a controlled drug delivery, since drug mobility in a solid lipid is lower compared with an oily phase. Other advantages of such carriers include the use of physiological compounds in the composition, the fast and effective production process, including the possibility of large scale production, the avoidance of organic solvents in the production procedures, and the possibility to produce high concentrated lipid suspensions [142]. The main disadvantage, however, is the low drug loading capacity [143], which is mainly related to the possibility of drug expulsion during storage [144].

Drug localization within SLN, as well as the capacity of these particles to retain the drug, will depend on the composition of the formulation (lipid, active compound, surfactant), as well as on the production conditions (hot vs. cold homogenization). The drug may be placed in between the chains of the fatty acids or in between the lipid layers and also in imperfections (e.g., amorphous clusters). After heating the melted lipids crystallize in higher energy conformations. These conformations are not very organized, allowing drugs to be loaded. During shelf life polymorphic transitions occur and low energy conformations are adopted, reducing the number of imperfections in the crystal lattice and expulsing the drug (Figure 4) [145,146]. Perfect lipid crystals are usually formed when the lipid molecules are chemically similar, e.g., pure triglycerides [147].

Nanostructured lipid carriers (NLC) are another type of lipid nanoparticles being developed to overcome some limitations of SLN. NLC are prepared not only from solid lipids but from a blend of a solid lipid with a certain amount of oil, to maintain a melting point above 40°C. Mixing especially very different molecules, such as long chain glycerides of the solid lipid with short chain glycerides of the liquid lipid, creates crystals with many imperfections [148]. Apart from localizing drug in between fatty acid chains or lipid lamellae, these imperfections provide a space for additional loading of active molecules. These latter can be incorporated in the particle matrix in a molecular dispersed form, or be arranged in amorphous clusters. There is also more flexibility for modulation of drug release, increasing the drug loading and preventing its leakage.

There are three models describing the NLC structure [148], namely, the imperfect type, the amorphous type, and the multiple type (Figure 4). Several comprehensive reviews have been devoted to this topic [144,148-152].

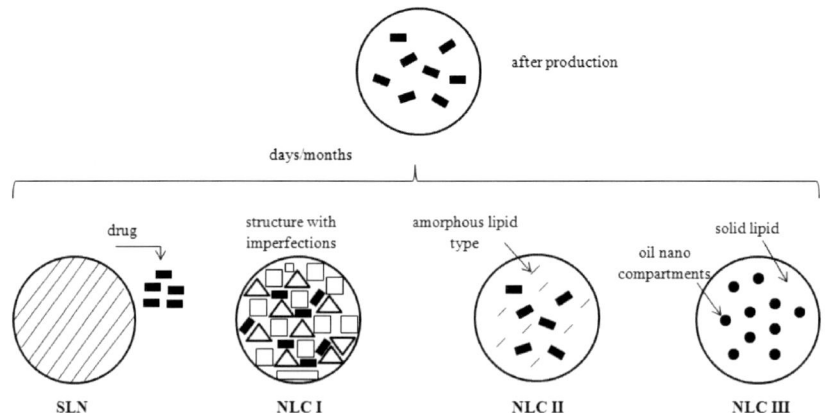

Figure 4. Crystallization process during storage to perfect crystal in SLN (left) and schematic model showing the NLC different structures: imperfect type (NLC I); amorphous type (NLCII) and multiple type (NLC III) (adapted from Müller et al. [148]).

Distribution of the drug depends on its physicochemical characteristics and on the composition of the particles, but it is influenced by the partition coefficient of the drug [153]. Lipid nanoparticles (SLN and NLC) are an interesting system for the ocular delivery of drugs. Similar to emulsions they are composed of accepted excipients, and can be produced on large industrial scale, using an established and low cost homogenization process. The lipid nanoparticles show additionally the advantages of a solid matrix similar to polymeric nanoparticles, having the ability to protect chemically labile ingredients and to modulate release (from very fast to extremely prolonged release). The possibility of surface modifications can be used to prolong pre-corneal residence time. Similarly to liposomes, several SLN and NLC have been successfully prepared for the incorporation of anti-inflammatory drugs but aimed for different administration routes, as intravenous, oral, pulmonary and transdermal. Examples of anti-inflammatory drug used include flurbiprofen [154-156], indomethacin [157-159], ketoprofen [158,160], naproxen [160], ketorolac [161], valdecoxib [162], sodium diclofenac [163], dexamethasone [164], hydrocortisone [163,165], betamethasone [166], triamcinolone acetonide [167,168], and methotrexate [169,170]. Therefore, it is expected that in the recent future lipid nanoparticles will become available for the treatment of ocular inflammatory diseases. Apart from the drug loading difficulties, several compounds commonly used in the treatment of ocular diseases have been incorporated in lipid nanoparticles, such as tobramycin [171], gatifloxacin [172], cyclosporine [173], and timolol maleate [174]. Lipid nanoparticles have shown sustained release and enhancement of drug bioavailability [171]. In the last years, SLN have attempted also a big development as non-viral systems for DNA delivery, but there are still few papers about their use in gene therapy. An example of this development is the work by del Pozo-Rodríguez et al. who evaluated the transfection capacity of SLN in the human retinal pigment epithelial established cell line in order to elucidate the potential application of this vector in the treatment of retinal diseases [175].

Lipid nanoparticles can be attractive systems for the ocular delivery of poorly water soluble drugs. For instance, hydrocortisone was loaded into SLN in a high extent (97% of total 0.5%, w/w). In vitro drug permeation studies through human cornea construct revealed that the permeation coefficients of hydrocortisone-loaded SLN were reduced in comparison to aqueous and oily solutions of the drug. However, the permeated amount was higher from the SLN due to a much higher hydrocortisone concentration achieved. The authors also observed that the high drug load of the nanoparticles provides prolonged drug release [176]. Sustained release and high permeation through bioengineered cornea was also observed for SLN containing diclofenac sodium. Using phospholipids together with homolipid from goat, the authors obtained high loading efficiencies (94%), which has been attributed it to complex formation between phospholipids and the drug, and also to drug association in the structures formed at the particle surface [177].

Ibuprofen has been widely used for the treatments of ocular inflammations. However, it is limited by a short biological half-life time (1.5–2 h) that leads to a short duration of action [178]. In order to overcome the shortcoming, multiple intakes of ibuprofen are required to maintain the effective concentration in human bodies which potentially lead to the occurrences of some side effects [179]. Due to its poor solubility in water, ibuprofen is an ideal candidate to be incorporated in lipid carriers. Since 2006 it has been successfully incorporate in SLN and NLC for diverse routes of delivery [178,180-184]. However, the first paper reporting a NLC intended for ocular drug delivery was published by Li et al. in 2008 [185]. They investigated four different compositions of NLC for the ocular delivery of ibuprofen. The effect of Gelucire 44/14 (as solid lipid material), Transcutol P (as permeability enhancer) and stearylamine (as charge-inducing reagent) on particle size, zeta potential, ocular irritation and corneal permeability were studied. The result showed that both Gelucire 44/14 and Transcutol P could enhance the drug corneal permeability to some extent; and stearylamine could prolong the pre-corneal retention of drug. Ibuprofen-loaded NLC displayed controlled-release profile. By optimizing the NLC composition, a suitable formulation was developed showing higher drug bioavailability in comparison to eye drops [185]. This prolonged pre-corneal retention has been attributed to the positively charged NLC.

The surface composition of the colloidal carriers may affect their affinity towards the ocular mucosa. The positive surface charge was found to have a positive effect in prolong the residence time of nanoemulsion droplet containing indomethacin on the epithelial layer of the cornea [94]. Nanoparticles coated with a bioadhesive polymer, such as hyaluronic acid or chitosan, indicated that the nature of the coating layers affect the interaction of the carriers with the corneal epithelial cells [186,187]. Based on these facts, a promising tool for ocular drug delivery is the use of NLC with surface modifications that could increase mucoadhesive properties, and leading to prolonged precorneal retention and, as a consequence, increased bioavailability. Such particles have recently been obtained by surface modification of cyclosporine loaded NLC with cysteine-polyethylene glycol stearate [188].

Surfactants applied in SLN for the treatment of ophthalmic diseases are selected from the group comprising: (i) lecithins, as they are, (e.g., Lipoids, phospholipids and hydrogenated forms thereof and synthetic and semi-synthetic derivates thereof); (ii) bile salts (e.g., sodium glycocholate, sodium taurocholate and taurodeoxycholate); (iii) polysorbates and sorbitans (e.g., Tween 20, Tween 40, Tween 80, Span 20, Span 40, and Span 60); and (iv) viscosity enhancers, particularly gelatin. The co-surfactants are selected from the group comprising: (i) low molecular weight alcohols or glycols (e.g., butanol, hexanol and hexadiol); (ii) low molecular weight fatty acids (e.g., butyric acid and octanoic acid); (iii) phosphoric acid esters, (iv) benzylic alcohol and (v) bile salts [189].

Several factors should be considered when choosing surfactants for lipid nanoparticles in general, namely, hydrophilic-lipophilic balance (HLB) value, route of administration, toxicity, and the effect on lipid modification and particle size. Surfactants with HLB values in the range of 8-18 are suitable for the preparation of o/w dispersion. Non-ionic surfactants are preferred over the ionic ones for ocular formulations because of their lower toxicity and irritancy. In general, the order of the toxicity of surfactants is from cationic to anionic, non-ionic, and ultimately amphoteric.

One important factor in the choice of a surfactant for lipid nanoparticles should be considered: different surfactants have variable influences on the in vivo biodegradation of the lipid matrix. For instance, non-ionic surfactants are more effective for inhibiting the degradation of lipid matrix in vivo. Finally, the choice of surfactants and co-surfactants also affects the particle size. Lipid nanoparticles made of the same lipids may have different sizes because of the use of different surfactants [190].

The development of lipid nanoparticles for ocular drug delivery has not received as much attention as other administration routes. However, some interesting properties highlighting their potential for ocular administration are their adhesiveness [191] and the longer retention time on the corneal surface and in the cul-de-sac, probably related to their relatively small size [171]. The nanoparticles are presumed to be entrapped and retained in the mucus layer [192].

Table 1. Aspects to be considered on choosing an ophthalmic delivery system and the performance of novel lipid-based systems

Aspects to consider	Novel lipid-based systems					
	Microemulsion	Nanoemulsion	Liposome	SLN	NLC
Facility of administration	+++	+++	++	++	++
Sensorial after administration – (avoidance of blurred vision, burning sensation)	+++	+++	++	++	++
Drug loading capacity	+++	+++	++	-	++
Possibility of drug targeting	-	-	+++	++	++
Pre-corneal retention time	++	++	++	++	++
In vitreous residence time	-	-	++	-*	-*
Controlled drug release	-	++	++	++	++
Avoidance of toxicity	-	++	+++	++	++
Scaling up of production	++	+++	-	+++	+++
Easy to sterilize	++	+++	-	+++	+++
Storage stability	+++	++	-	++	++

+ + + excellent, + + very good, + good, - poor, -*not reported.

Table 2. Advantages and disadvantages of novel lipid-based drug delivery systems and anti-inflammatory drugs incorporated for ophthalmic delivery

Lipid based delivery system	Advantages	Disadvantages	Anti-inflammatory drugs used for ocular delivery
Microemulsion	High solubility of both hydrophilic and hydrophobic drugs Easy preparation and sterilization Prolonged pre-corneal residence time Enhanced drug penetration Transparent systems	Large amounts of surfactants and co-surfactants needed Potential toxicity	Dexamethasone [194]
Nanoemulsions	Small amount of surfactant – toxicological safety High solubility of both hydrophilic and hydrophobic drugs Easy preparation – large scale production Easy sterilization May be transparent Prolonged pre-corneal residence time	Difficult to control drug release Thermodynamically instable systems	Indomethacin [80, 89, 90, 94, 95] Piroxicam [93] Difluprednate [81] Predinisolone [84]
Liposomes	Biocompatible and biodegradable Able to entrap both hydrophilic and hydrophobic drugs Controlled release Provide drug protection from metabolic degradation Prolonged residence time – pre-corneal and in vitreous	Poor stability Difficult to prepare and sterilize High cost	Diclofenac sodium [127, 128] Dexamethasone [124-126, 195]
SLN	Easy preparation – large scale production Easy sterilization Improved ocular bioavailability Prolonged pre-corneal residence time Controlled release	Limited drug loading	Hydrocortisone [176] Diclofenac sodium[177]

Table 2. (Continued).

Lipid based delivery system	Advantages	Disadvantages	Anti-inflammatory drugs used for ocular delivery
NLC	Easy preparation – large scale production Easy sterilization Drug loading of lipophilic and possibly hydrophilic Improved ocular bioavailability Prolonged pre-corneal residence time Controlled release	Hydrophilic drugs can show burst effects	Ibuprofen [185]

Table 3. List of surfactants, co-surfactants commonly used in SLN and NLC

Amphoteric surfactants	Non-ionic surfactants	Ionic surfactants	Co-surfactants
Egg phosphatidylcholine Egg lecithin Soy phosphatidylcholine (Epikuron 200, 95% SP) (Lipoid S100) (Lipoid S75, 68% SP) (Lipoid S75, 68% SP) (Phospholipon 90G, 90%)	Poloxamer 188 Poloxamer 207 Poloxamine 908 Polyglycerol methyl glucose distearate Solutol HS15 Span 20, 40, 60 Trehalose Tween 20, 40, 60 Tyloxapol	Sodium cholate Sodium dodecyl sulfate Sodium glycocholate Sodium oleate Sodium taurocholate Sodium taurodeoxycholate	Butanol Butyric acid

Table 4. Comparison between SLN/NLC and polymeric nanoparticles (adapted from Date et al. [193])

	SLN/NLC	Polymeric nanoparticles
Ocular delivery	Possible	Possible
Ability to deliver hydrophobic and hydrophilic drugs	Yes	Yes
Physical stability	+++	+++
Biological stability	++	+++
Biocompatibility	+++	++
Ease of sterilization	++	++
Drug targeting	++	++
Drug loading	Low to moderate (SLN), moderate to high (NLC)	Moderate
Ease of commercialization	++	+
Ability to deliver biotechnological therapeutics	++	++

Chapter IV

Conclusions and Future Trends

For the treatment of ocular inflammatory diseases, one should keep in mind that there are not ideal anti-inflammatory agents or administration regimens. As such, the pharmacological treatment should be chosen considering drug mechanism of action, disease specific conditions, possible side effects and the drug ability on reaching the site of infection in therapeutic concentrations. Although microemulsions present excellent advantages over conventional systems, as high drug solubilization, long stability and enhanced corneal penetration, they suffer from limitations in selection of surfactants/co-surfactants due to its potential toxicity. On the contrary, nanoemulsions can be easily prepared with less amount of surfactants, being in this way well tolerated. They can be used to increase pre-corneal residence time as well as penetration of drugs, though its major drawback is the limited ability on sustaining the drug release. Liposomes and SLN could be therefore a suitable alternative. However, the former represents a challenge when considering large scale production, whereas the latter depicts a low drug loading capacity. NLC have emerged as a novel delivery system that could incorporate the advantages of the lipid based delivery systems and overcome their limitations. In the last few years, several NLC formulations have been successfully prepared for the incorporation of anti-inflammatory drugs but have not yet been fully employed in the treatment of ocular diseases. It is expected that in the next years more studies will be performed using lipid based systems for the ocular route, resulting, in this way, in more efficient therapeutic options for the treatment of inflammatory diseases.

References

[1] Ahuja, M.; Dhake, A.; Sharma, S.; Majumdar, D., Topical ocular delivery of NSAIDs. *AAPS J.* 2008, 10, (2), 229-241.
[2] Elhers, W. H.; Donshik, P. C., Giant papillary conjunctivitis. *Curr. Opin. Allergy Clin. Immunol.* 2008, 8, (5), 445-9.
[3] Leonardi, A.; Motterle, L.; Bortolotti, M., Allergy and the eye. *Clin. Exp. Immunol.* 2008, 153 Suppl 1, 17-21.
[4] Bielory, L.; Friedlaender, M. H., Allergic conjunctivitis. *Immunol. Allergy Clin. North Am.* 2008, 28, (1), 43-58.
[5] Yeh, S.; Faia, L. J.; Nussenblatt, R. B., Advances in the diagnosis and immunotherapy for ocular inflammatory disease. *Semin. Immunopathol.* 2008, 30, (2), 145-64.
[6] Pflugfelder, S. C., Antiinflammatory therapy for dry eye. *Am. J. Ophthalmol.* 2004, 137, (2), 337-42.
[7] Struck, H. G.; Bariszlovich, A., Comparison of 0.1% dexamethasone phosphate eye gel (Dexagel) and 1% prednisolone acetate eye suspension in the treatment of post-operative inflammation after cataract surgery. *Graefes Archive for Clinical and Experimental Ophthalmology.* 2001, 239, (10), 737-742.
[8] Hoang-Xuan, T.; Hannouche, D., Medical treatment of ocular burns. *Journal Francais D. Ophtalmologie.* 2004, 27, (10), 1175-1178.
[9] Pavesio, C. E.; Decory, H. H., Treatment of ocular inflammatory conditions with loteprednol etabonate. *Br. J. Ophthalmol.* 2008, 92, (4), 455-459.
[10] Abad, S.; Seve, P.; Dhote, R.; Brezin, A. P., Guidelines for the management of uveitis in Internal Medicine. *Revue de Medecine Interne.* 2009, 30, (6), 492-500.

[11] Messmer, E. M., Therapeutic options in vernal keratoconjunctivitis. *Ophthalmologe.* 2009, 106, (6), 557-561.
[12] Group, L. E. U. U. S., Controlled evaluation of loteprednol etabonate and prednisolone acetate in the treatment of acute anterior uveitis. Loteprednol Etabonate US Uveitis Study Group. *Am. J. Ophthalmol.* 1999, 127, (5), 537-544.
[13] Raizman, M., Corticosteroid therapy of eye disease. Fifty years later. *Arch. Ophthalmol.* 1996, 114, (8), 1000-1001.
[14] McGhee, C. N.; Dean, S.; Danesh-Meyer, H., Locally administered ocular corticosteroids: benefits and risks. *Drug Saf.* 2002, 25, (1), 33-55.
[15] Psilas, K.; Kalogeropoulos, C.; Loucatzicos, E.; Asproudis, I.; Petroutsos, G., The Effect of Indomethacin, Diclofenac and Flurbiprofen on the Maintenance of Mydriasis During Extracapsular Cataract-Extraction. *Documenta Ophthalmologica.* 1992, 81, (3), 293-300.
[16] Stewart, R.; Grosserode, R.; Cheetham, J. K.; Rosenthal, A., Efficacy and safety profile of ketorolac 0.5% ophthalmic solution in the prevention of surgically induced miosis during cataract surgery. *Clin. Ther.* 1999, 21, (4), 723-732.
[17] Russo, P.; Papa, V.; Russo, S.; Di Bella, A.; Pabst, G.; Milazzo, G.; Balestrazzi, A.; Caporossi, A., Topical nonsteroidal anti-inflammatory drugs in uncomplicated cataract surgery: Effect of sodium naproxen. *Eur. J. Ophthalmol.* 2005, 15, (5), 598-606.
[18] Swamy, B. N.; Chilov, M.; McClellan, K.; Petsoglou, C., Topical non-steroidal anti-inflammatory drugs in allergic conjunctivitis: Meta-analysis of randomized trial data. *Ophthalmic Epidemiol.* 2007, 14, (5), 311-319.
[19] Yavas, G. F.; Ozturk, F.; Kusbeci, T., Preoperative topical indomethacin to prevent pseudophakic cystoid macular edema. *J. Cataract Refract. Surg.* 2007, 33, (5), 804-807.
[20] Warren, K. A.; Fox, J. E., Topical Nepafenac As An Alternate Treatment for Cystoid Macular Edema in Steroid Responsive Patients. *Retina-the Journal of Retinal and Vitreous Diseases.* 2008, 28, (10), 1427-1434.
[21] Aslam, S. A.; Sheth, H. G.; Vaughan, A. J., Emergency management of corneal injuries. *Injury-International Journal of the Care of the Injured.* 2007, 38, (5), 594-597.
[22] Wilson, S. A.; Last, A., Management of corneal abrasions. *Am. Fam. Physician.* 2004, 70, (1), 123-128.
[23] Sun, R.; Gimbel, H. V., Effects of topical ketorolac and diclofenac on normal corneal sensation. *J. Refract. Surg.* 1997, 13, (2), 158-161.

[24] Shimazaki, J.; Saito, H.; Yang, H. Y.; Toda, I.; Fujishima, H.; Tsubota, K., Persistent epithelial defect following penetrating keratoplasty: an adverse effect of diclofenac eyedrops. *Cornea.* 1995, 14, (6), 623-627.
[25] Gills, J. P., Voltaren associated with medication keratitis. *J. Cataract. Refract. Surg.* 1994, 20, (1), 110.
[26] Barar, J.; Javadzadeh, A. R.; Omidi, Y., Ocular novel drug delivery: impacts of membranes and barriers. *Expert Opin. Drug Deliv.* 2008, 5, (5), 567-581.
[27] Schalenbourg, A.; Leys, A.; de, C.; Coutteel, C.; Herbort, C. P., Corticosteroid-induced central serous chorioretinopathy in patients with ocular inflammatory disorders. *Klinische Monatsblatter fur Augenheilkunde.* 2002, 219, (4), 264-267.
[28] Cunningham, M. A.; Edelman, J. L.; Kaushal, S., Intravitreal steroids for macular edema: The past, the present, and the future. *Surv. Ophthalmol.* 2008, 53, (2), 139-149.
[29] Del Amo, E. M.; Urtti, A., Current and future ophthalmic drug delivery systems. A shift to the posterior segment. *Drug Discov. Today.* 2008, 13, (3-4), 135-143.
[30] Moshfeghi, D. M.; Kaiser, P. K.; Scott, I. U.; Sears, J. E.; Benz, M.; Sinesterra, J. P.; Kaiser, R. S.; Bakri, S. J.; Maturi, R. K.; Belmont, J.; Beer, P. M.; Murray, T. G.; Quiroz-Mercado, H.; Mieler, W. F., Acute endophthalmitis following intravitreal triamcinolone acetonide injection. *Am. J. Ophthalmol.* 2003, 136, (5), 791-796.
[31] Fialho, S. L.; Siqueira, R. C.; Jorge, R.; Silva-Cunha, A., Biodegradable implants for ocular delivery of anti-inflammatory drug. *J. Drug Deliv. Sci. Technol.* 2007, 17, (1), 93-97.
[32] Okabe, K.; Kimura, H.; Okabe, J.; Ogura, Y., Ocular tissue distribution of betamethasone after anterior-episcleral, posterior-episcleral, and anterior-intrascleral placement of nonbiodegradable implants. *Retina-the Journal of Retinal and Vitreous Diseases.* 2007, 27, (6), 770-777.
[33] Taban, M.; Lowder, C. Y.; Kaiser, P. K., Outcome of Fluocinolone Acetonide Implant (Retisert (Tm)) Reimplantation for Chronic Noninfectious Posterior Uveitis. *Retina-the Journal of Retinal and Vitreous Diseases.* 2008, 28, (9), 1280-1288.
[34] Wolf, E. J.; Braunstein, A.; Shih, C.; Braunstein, R. E., Incidence of visually significant pseudophakic macular edema after uneventful phacoemulsification in patients treated with nepafenac. *J. Cataract. Refract. Surg.* 2007, 33, (9), 1546-1549.

[35] Labetoulle, M.; Frau, E.; Le Jeunne, C., Systemic adverse effects of topical ocular treatments. *Presse Med.* 2005, 34, (8), 589-595.
[36] Schalnus, R., Topical nonsteroidal anti-inflammatory therapy in ophthalmology. *Ophthalmologica.* 2003, 217, (2), 89-98.
[37] Awan, M. A.; Agarwal, P. K.; Watson, D. G.; McGhee, C. N. J.; Dutton, G. N., Penetration of topical and subconjunctival corticosteroids into human aqueous humour and its therapeutic significance. *Br. J. Ophthalmol.* 2009, 93, (6), 708-713.
[38] Anumolu, S. S.; Singh, Y.; Gao, D.; Stein, S.; Sinko, P. J., Design and evaluation of novel fast forming pilocarpine-loaded ocular hydrogels for sustained pharmacological response. *J. Control Releaseease.* 2009, 137, (2), 152-159.
[39] Bucolo, C.; Spadaro, A., Pharmacological evaluation of anti-inflammatory pyrrole-acetic acid derivative eye drops. *J. Ocul. Pharmacol. Ther.* 1997, 13, (4), 353-361.
[40] Adibkia, K.; Shadbad, M. R. S.; Nokhodchi, A.; Javadzedeh, A.; Barzegar-Jalali, M.; Barar, J.; Mohammadi, G.; Omidi, Y., Piroxicam nanoparticles for ocular delivery: Physicochemical characterization and implementation in endotoxin-induced uveitis. *J. Drug Target.* 2007, 15, (6), 407-416.
[41] Gupta, A. K.; Madan, S.; Majumdar, D. K.; Maitra, A., Ketorolac entrapped in polymeric micelles: preparation, characterisation and ocular anti-inflammatory studies. *Int. J. Pharm.* 2000, 209, (1-2), 1-14.
[42] Maxwell, W. A.; Reiser, H. J.; Stewart, R. H.; Cavanagh, H. D.; Walters, T. R.; Sager, D. P.; Meuse, P. A., Nepafenac Dosing Frequency for Ocular Pain and Inflammation Associated with Cataract Surgery. *J. Ocul. Pharmacol. Ther.* 2008, 24, (6), 593-599.
[43] Shirasaki, Y., Molecular design for enhancement of ocular penetration. *J. Pharm. Sci.* 2008, 97, (7), 2462-2496.
[44] Unger, W. G., Review: mediation of the ocular response to injury. *J. Ocul. Pharmacol.* 1990, 6, (4), 337-53.
[45] Acosta, M. C.; Luna, C.; Graff, G.; Meseguer, V. M.; Viana, F.; Gallar, J.; Belmonte, C., Comparative effects of the nonsteroidal anti-inflammatory drug nepafenac on corneal sensory nerve fibers responding to chemical irritation. *Invest. Ophthalmol. Vis. Sci.* 2007, 48, (1), 182-188.
[46] Stein, R.; Stein, H. A.; Cheskes, A.; Symons, S., Photorefractive keratectomy and postoperative pain. *Am. J. Ophthalmol.* 1994, 117, (3), 403-5.

[47] Heier, J. S.; Topping, T. M.; Baumann, W.; Dirks, M. S.; Chern, S., Ketorolac versus prednisolone versus combination therapy in the treatment of acute pseudophakic cystoid macular edema. *Ophthalmology.* 2000, 107, (11), 2034-2038.

[48] Jaanus, S. D.; Lesher, G. A., Anti-inflammatory drugs. In *Clinical ocular pharmacology*, Bartlett J.D., J. S. D., Ed. Butterworth-Heineman: Boston, 1995; pp 303-314.

[49] Campbell, W. B.; Halushka, P. V., Lipid derived autocoids. In *Goodman and Gilman's. The Pharmacological Basis of Therapeutics. 9th ed*, Hardman, J. G. a. L., L.E., Ed. McGraw-Hill: New York, 1990; pp 601-33.

[50] Gaudio, P., A review of evidence guiding the use of corticosteroids in the treatment of intraocular inflammation. *Ocular Immunol. Inflamm.* 2004, 12, 169-192.

[51] Jaffe, G. J.; Ben-Nun, J.; Guo, H.; Dunn, J. P.; Ashton, P., Fluocinolone acetonide sustained drug delivery device to treat severe uveitis. *Ophthalmology.* 2000, 107, (11), 2024-33.

[52] Jaffe, G. J.; Pearson, P. A.; Ashton, P., Dexamethasone sustained drug delivery implant for the treatment of severe uveitis. *Retina.* 2000, 20, (4), 402-3.

[53] Samudre, S. S.; Lattanzio, F. A., Jr.; Williams, P. B.; Sheppard, J. D., Jr., Comparison of topical steroids for acute anterior uveitis. *J. Ocul. Pharmacol. Ther.* 2004, 20, (6), 533-47.

[54] Recasens, J. F.; Green, K., Effects of endotoxin and anti-inflammatory agents on superoxide dismutase in the rabbit iris. *Ophthalmic Res.* 1990, 22, (1), 12-8.

[55] Bron, A.; Denis, P.; T.C., H. X.; P., C.; Hachet, E.; Medhorn, E.; Akingbehin, A., The effects of Rimexolone 1% in postoperative inflammation after cataract extraction. A double-masked placebo-controlled study. *Eur. J. Ophthalmol.* 1998, 8, 16 -21.

[56] Nagelhout, T. J.; Gamache, D. A.; Roberts, L.; Brady, M. T.; Yanni, J. M., Preservation of tear film integrity and inhibition of corneal injury by dexamethasone in a rabbit model of lacrimal gland inflammation-induced dry eye. *J. Ocul. Pharmacol. Ther.* 2005, 21, (2), 139-48.

[57] Renfro, L.; Snow, J. S., Ocular effects of topical and systemic steroids. *Dermatol. Clin.* 1992, 10, (3), 505-12.

[58] O'Brien, T. P., Emerging guidelines for use of NSAID therapy to optimize cataract surgery patient care. *Curr. Med. Res. Opin.* 2005, 21, (7), 1131-7.

[59] Reddy, M. S.; Suneetha, N.; Thomas, R. K.; Battu, R. R., Topical diclofenac sodium for treatment of postoperative inflammation in cataract surgery. *Indian J. Ophthalmol.* 2000, 48, (3), 223-6.

[60] Flach, A. J., Topical nonsteroidal antiinflammatory drugs in ophthalmology. *Int. Ophthalmol. Clin.* 2002, 42, (1), 1-11.

[61] Leonardi, A.; Borghesan, F.; Faggian, D.; Depaoli, M.; Secchi, A. G.; Plebani, M., Tear and serum soluble leukocyte activation markers in conjuctival allergic diseases. *Am. J. Ophthalmol.* 2000, 129, (2), 151-8.

[62] Miyake, K.; Ibaraki, N., Prostaglandins and cystoid macular edema. *Surv. Ophthalmol.* 2002, 47(Suppl.1), 203-218.

[63] Price, M. O.; Price, F. W., Efficacy of topical ketorolac tromethamine 0.4% for control of pain or discomfort associated with cataract surgery. *Curr. Med. Res. Opin.* 2004, 20, (12), 2015-9.

[64] Bandare, B. M.; Sankaridurg, P. R.; Willcox, M. D., Non-steroidal antinflammatory agents decrease bacterial colonization of contact lenses and prevent adhesion to human corneal epithelial cells. *Curr. Eye Res.* 2004, 29, 245-251.

[65] Walters, T.; Raizman, M.; Ernest, P.; Gayton, J.; Lehmann, R., In vivo pharmacokinetics and in vitro pharmacodynamics of nepafenac, amfenac, ketorolac, and bromfenac. *J. Cataract Refract. Surg.* 2007, 33, (9), 1539-45.

[66] Brauninger, G. E.; Shah, D. O.; Kaufman, H. E., Direct physical demonstration of oily layer on tear film surface. *Am. J. Ophthalmol.* 1972, 73, (1), 132-134.

[67] Malhotra, M.; Majumdar, D. K., In vivo ocular availability of ketorolac following ocular instillations of aqueous, oil, and ointment formulations to normal corneas of rabbits: A technical note. *Aaps. Pharmscitech.* 2005, 6, (3).

[68] Nanjawade, B. K.; Manvi, F. V.; Manjappa, A. S., In situ.-forming hydrogels for sustained ophthalmic drug delivery. *J. Control Release.* 2007, 122, (2), 119-134.

[69] Hoar, T. P.; Schulman, J. H., Transparent water-in-oil dispersions the oleopathic hydro-micelle. *Nature.* 1943, 152, 102-103.

[70] Lawrence, M. J.; Rees, G. D., Microemulsion-based media as novel drug delivery systems. *Adv. Drug Deliv. Rev.* 2000, 45, (1), 89-121.

[71] Gupta, S.; Moulik, S. P., Biocompatible microemulsions and their prospective uses in drug delivery. *J. Pharm. Sci.* 2008, 97, (1), 22-45.

[72] Alany, R. G.; Rades, T.; Nicoll, J.; Tucker, I. G.; Davies, N. M., W/O microemulsions for ocular delivery: Evaluation of ocular irritation and precorneal retention. *J. Control Release.* 2006, 111, (1-2), 145-152.

[73] Bonacucina, G.; Cespi, M.; Misici-Falzi, M.; Palmieri, G. F., Colloidal soft matter as drug delivery system. *J. Pharm. Sci.* 2009, 98, (1), 1-42.

[74] Siebenbrodt, I.; Keipert, S., Poloxamer-Systems As Potential Ophthalmics. 2. Microemulsions. *Eur. J. Pharm. Biopharm.* 1993, 39, (1), 25-30.

[75] Hasse, A.; Keipert, S., Development and characterization of microemulsions for ocular application. *Eur. J. Pharm. Biopharm.* 1997, 43, (2), 179-183.

[76] Fialho, S. L.; Silva-Cunha, A., New vehicle based on a microemulsion for topical ocular administration of dexamethasone. *Clin. Experiment. Ophthalmol.* 2004, 32, (6), 626-632.

[77] Gaudana, R.; Jwala, J.; Boddu, S. H.; Mitra, A. K., Recent perspectives in ocular drug delivery. *Pharm. Res.* 2009, 26, (5), 1197-1216.

[78] El Aasser, M. S.; Sudol, E. D., Miniemulsions: Overview of research and applications. *Jct. Research.* 2004, 1, (1), 21-31.

[79] Muchtar, S.; Abdulrazik, M.; FruchtPery, J.; Benita, S., Ex-vivo permeation study of indomethacin from a submicron emulsion through albino rabbit cornea. *J. Control Release.* 1997, 44, (1), 55-64.

[80] Badawi, A. A.; El-Laithy, H. M.; El Qidra, R. K.; El Mofty, H.; El dally, M., Chitosan based nanocarriers for indomethacin ocular delivery. *Arch. Pharm. Res.* 2008, 31, (8), 1040-9.

[81] Yamaguchi, M.; Yasueda, S.; Isowaki, A.; Yamamoto, M.; Kimura, M.; Inada, K.; Ohtori, A., Formulation of an ophthalmic lipid emulsion containing an anti-inflammatory steroidal drug, difluprednate. *Int. J. Pharm.* 2005, 301, (1-2), 121-8.

[82] Tamilvanan, S., Oil-in-water lipid emulsions: implications for parenteral and ocular delivering systems. *Progress in Lipid Research.* 2004, 43, (6), 489-533.

[83] Tamilvanan, S.; Benita, S., The potential of lipid emulsion for ocular delivery of lipophilic drugs. *Eur. J. Pharm. Biopharm.* 2004, 58, (2), 357-368.

[84] Ibrahim, S. S.; Awad, G. A.; Geneidi, A.; Mortada, N. D., Comparative effects of different cosurfactants on sterile prednisolone acetate ocular submicron emulsions stability and release. *Colloids Surf. B. Biointerfaces.* 2009, 69, (2), 225-31.

[85] Gutierrez, J. M.; Gonzalez, C.; Maestro, A.; Sole, I.; Pey, C. M.; Nolla, J., Nano-emulsions: New applications and optimization of their preparation. *Curr. Opin. Colloid Interface Sci.* 2008, 13, (4), 245-251.

[86] Solans, C.; Izquierdo, P.; Nolla, J.; Azemar, N.; Garcia-Celma, M. J., Nano-emulsions. *Curr. Opin. Colloid Interface Sci.* 2005, 10, (3-4), 102-110.

[87] Beilin, M.; Barilan, A.; Amselem, S.; Schwarz, J.; Yogev, A.; Neumann, R., Ocular Retention Time of Submicron Emulsion (Sme) and the Miotic Response to Pilocarpine Delivered in Sme. *Invest. Ophthalmol. Vis. Sci.* 1995, 36, (4), S166-S166.

[88] Mehnert, W.; Mader, K., Solid lipid nanoparticles: production, characterization and applications. *Adv. Drug Deliv. Rev.* 2001, 47, (2-3), 165-196.

[89] Calvo, P.; Vila-Jato, J. L.; Alonso, M. J., Comparative in vitro evaluation of several colloidal systems, nanoparticles, nanocapsules, and nanoemulsions, as ocular drug carriers. *J. Pharm. Sci.* 1996, 85, (5), 530-6.

[90] Calvo, P.; Alonso, M. J.; Vila-Jato, J. L.; Robinson, J. R., Improved ocular bioavailability of indomethacin by novel ocular drug carriers. *J. Pharm. Pharmacol.* 1996, 48, (11), 1147-52.

[91] Gershkovich, P.; Wasan, K. M.; Barta, C. A., A Review of the Application of Lipid-Based Systems in Systemic, Dermal/Transdermal, and Ocular Drug Delivery. *Crit. Rev. Ther. Drug Carrier Syst.* 2008, 25, (6), 545-584.

[92] Rabinovich-Guilatt, L.; Couvreur, P.; Lambert, G.; Dubernet, C., Cationic vectors in ocular drug delivery. *J. Drug Target.* 2004, 12, (9-10), 623-633.

[93] Klang, S. H.; Siganos, C. S.; Benita, S.; Frucht-Pery, J., Evaluation of a positively charged submicron emulsion of piroxicam on the rabbit corneum healing process following alkali burn. *J. Control Release.* 1999, 57, (1), 19-27.

[94] Klang, S.; Abdulrazik, M.; Benita, S., Influence of emulsion droplet surface charge on indomethacin ocular tissue distribution. *Pharm. Dev. Technol.* 2000, 5, (4), 521-32.

[95] Yamaguchi, M.; Ueda, K.; Isowaki, A.; Ohtori, A.; Takeuchi, H.; Ohguro, N.; Tojo, K., Mucoadhesive properties of chitosan-coated ophthalmic lipid emulsion containing indomethacin in tear fluid. *Biol. Pharm. Bull.* 2009, 32, (7), 1266-71.

[96] Meisner, D.; Mezei, M., Liposome Ocular Delivery Systems. *Adv. Drug Deliv. Rev.* 1995, 16, (1), 75-93.
[97] Bangham, A. D.; Horne, R. W., Negative staining of phospholipids and their structural modification by surface-active agents as observed in the electron microscope. *J Mol Biol* 1964, 8, 660-668.
[98] Ebrahim, S.; Peyman, G. A.; Lee, P. J., Applications of liposomes in ophthalmology. *Surv Ophthalmol* 2005, 50, (2), 167-182.
[99] Mehanna, M. M.; Elmaradny, H. A.; Samaha, M. W., Ciprofloxacin liposomes as vesicular reservoirs for ocular delivery: formulation, optimization, and in vitro characterization. *Drug Dev Ind Pharm* 2009, 35, (5), 583-593.
[100] Bejjani, R. A.; Jeanny, J. C.; Bochot, A.; Behar-Cohen, F., The use of liposomes as intravitreal drug delivery system. *Journal Francais D Ophtalmologie.* 2003, 26, (9), 981-985.
[101] Barza, M.; Stuart, M.; Szoka, F., Jr., Effect of size and lipid composition on the pharmacokinetics of intravitreal liposomes. *Invest. Ophthalmol. Vis. Sci.* 1987, 28, (5), 893-900.
[102] Liu, K. R.; Peyman, G. A.; Khoobehi, B.; Alkan, H.; Fiscella, R., Intravitreal liposome-encapsulated trifluorothymidine in a rabbit model. *Ophthalmology.* 1987, 94, (9), 1155-1159.
[103] Alghadyan, A. A.; Peyman, G. A.; Khoobehi, B.; Milner, S.; Liu, K. R., Liposome-bound cyclosporine: clearance after intravitreal injection. *Int. Ophthalmol.* 1988, 12, (2), 109-112.
[104] Kawakami, S.; Yamamura, K.; Mukai, T.; Nishida, K.; Nakamura, J.; Sakaeda, T.; Nakashima, M.; Sasaki, H., Sustained ocular delivery of tilisolol to rabbits after topical administration or intravitreal injection of lipophilic prodrug incorporated in liposomes. *J. Pharm. Pharmacol.* 2001, 53, (8), 1157-1161.
[105] Zeng, S.; Hu, C. Z.; Wei, H. R.; Lu, Y. S.; Zhang, Y.; Yang, J. X.; Yun, G. X.; Zou, W. P.; Song, B. P., Intravitreal Pharmacokinetics of Liposome-Encapsulated Amikacin in A Rabbit Model. *Ophthalmology.* 1993, 100, (11), 1640-1644.
[106] Wiechens, B.; Grammer, J. B.; Johannsen, U.; Pleyer, U.; Hedderich, J.; Duncker, G. I. W., Experimental intravitreal application of ciprofloxacin in rabbits. *Ophthalmologica.* 1999, 213, (2), 120-128.
[107] Wiechens, B.; Krausse, R.; Grammer, J. B.; Neumann, D.; Pleyer, U.; Duncker, G. I. W., Clearance of liposome-incorporated ciprofloxacin after intravitreal injection in rabbit eyes. *Klinische Monatsblatter fur Augenheilkunde.* 1998, 213, (5), 284-292.

[108] Wiechens, B.; Neumann, D.; Grammer, J. B.; Pleyer, U.; Hedderich, J.; Duncker, G. I. W., Retinal toxicity of liposome-incorporated and free ofloxacin after intravitreal injection in rabbit eyes. *Int. Ophthalmol.* 1998, 22, (3), 133-143.

[109] Wiechens, B.; Schutze, D.; Grammer, J. B.; Krause, R.; Pleyer, U.; Duncker, G., Clearance of free and liposome-incorporated Norfloxacin after intravitreal injection. *Invest. Ophthalmol. Vis. Sci.* 1999, 40, (4), S87-S87.

[110] Cheng, C. K.; Yang, C. H.; Hsueh, P. R.; Liu, C. M.; Lu, H. Y., Vitrectomy with fluconazole infusion: Retinal toxicity, pharmacokinetics, and efficacy in the treatment of experimental candidal endophthalmitis. *J. Ocul. Pharmacol. Ther.* 2004, 20, (5), 430-438.

[111] Taskintuna, I.; Rahhal, F. M.; Arevalo, J. F.; Munguia, D.; Banker, A. S.; DeClercq, E.; Freeman, W. R., Low-dose intravitreal cidofovir (HPMPC) therapy of cytomegalovirus retinitis in patients with acquired immune deficiency syndrome. *Ophthalmology.* 1997, 104, (6), 1049-1057.

[112] Cheng, L. Y.; Hostetler, K. Y.; Ozerdem, U.; Gardner, M. F.; Mach-Hofacre, B.; Freman, W. R., Intravitreal toxicology and treatment efficacy of a long acting anti-viral lipid prodrug of ganciclovir (HDP-GCV) in liposome formulation. *Invest. Ophthalmol. Vis. Sci.* 1999, 40, (4), S872-S872.

[113] Cheng, L. Y.; Hostetler, K. Y.; Toyoguchi, M.; Beadle, J. R.; Rodanant, N.; Gardner, M. F.; Aldern, K. A.; Bergeron-Lynn, G.; Freeman, W. R., Ganciclovir release rates in vitreous from different formulations of 1-O-hexadecylpropanediol-3-phospho-ganciclovir. *J. Ocul. Pharmacol. Ther.* 2003, 19, (2), 161-169.

[114] Yasukawa, T.; Kimura, H.; Kunou, N.; Miyamoto, H.; Honda, Y.; Ogura, Y.; Ikada, Y., Biodegradable scleral implant for intravitreal controlled release of ganciclovir. *Graefes Archive for Clinical and Experimental Ophthalmology.* 2000, 238, (2), 186-190.

[115] Lallemand, F.; Felt-Baeyens, O.; Besseghir, K.; Behar-Cohen, F.; Gurny, R., Cyclosporine A delivery to the eye: a pharmaceutical challenge. *Eur. J. Pharm. Biopharm.* 2003, 56, (3), 307-318.

[116] Kawakami, S.; Harada, A.; Sakanaka, K.; Nishida, K.; Nakamura, J.; Sakaeda, T.; Ichikawa, N.; Nakashima, M.; Sasaki, H., In vivo gene transfection via intravitreal injection of cationic liposome/plasmid DNA complexes in rabbits. *Int. J. Pharm.* 2004, 278, (2), 255-262.

[117] Peeters, L.; Sanders, N. N.; Jones, A.; Demeester, J.; De Smedt, S. C., Post-pegylated lipoplexes are promising vehicles for gene delivery in RPE cells. *J. Control Release.* 2007, 121, (3), 208-217.

[118] Arakawa, Y.; Hashida, N.; Ohguro, N.; Yamazaki, N.; Onda, M.; Matsumoto, S.; Ohishi, M.; Yamabe, K.; Tano, Y.; Kurokawa, N., Eye-concentrated distribution of dexamethasone carried by sugar-chain modified liposome in experimental autoimmune uveoretinitis mice. *Biomedical Research-Tokyo.* 2007, 28, (6), 331-334.

[119] Schaeffer, H. E.; Breitfeller, J. M.; Krohn, D. L., Lectin-Mediated Attachment of Liposomes to Cornea - Influence on Trans-Corneal Drug Flux. *Invest. Ophthalmol. Vis. Sci.* 1982, 23, (4), 530-533.

[120] McCalden, T. A.; Levy, M., Retention of Topical Liposomal Formulations on the Cornea. *Experientia.* 1990, 46, (7), 713-715.

[121] Monem, A. S.; Ali, F. M.; Ismail, M. W., Prolonged effect of liposomes encapsulating pilocarpine HCl in normal and glaucomatous rabbits. *Int. J. Pharm.* 2000, 198, (1), 29-38.

[122] Mainardes, R. M.; Urban, M. C. C.; Cinto, P. O.; Khalil, N. M.; Chaud, M. V.; Evangelista, R. C.; Gremiao, M. P. D., Colloidal carriers for ophthalmic drug delivery. *Curr. Drug Targets.* 2005, 6, (3), 363-371.

[123] Singh, K.; Mezei, M., Liposomal Ophthalmic Drug Delivery System. 1. Triamcinolone Acetonide. *Int. J. Pharm.* 1983, 16, (3), 339-344.

[124] Taniguchi, K.; Itakura, K.; Yamazawa, N.; Morisaki, K.; Hayashi, S.; Yamada, Y., Efficacy of a liposome preparation of anti-inflammatory steroid as an ocular drug-delivery system. *J. Pharmacobiodyn.* 1988, 11, (1), 39-46.

[125] Al-Muhammad, J.; Ozer, A. Y.; Hincal, A. A., Studies on the formulation and in vitro release of ophthalmic liposomes containing dexamethasone sodium phosphate. *J. Microencapsul.* 1996, 13, (2), 123-30.

[126] Al-Muhammed, J.; Ozer, A. Y.; Ercan, M. T.; Hincal, A. A., In-vivo studies on dexamethasone sodium phosphate liposomes. *J. Microencapsul.* 1996, 13, (3), 293-306.

[127] Sun, K. X.; Wang, A. P.; Huang, L. J.; Liang, R. C.; Liu, K., [Preparation of diclofenac sodium liposomes and its ocular pharmacokinetics]. *Yao Xue Xue Bao.* 2006, 41, (11), 1094-8.

[128] Li, N.; Zhuang, C.; Wang, M.; Sun, X.; Nie, S.; Pan, W., Liposome coated with low molecular weight chitosan and its potential use in ocular drug delivery. *Int. J. Pharm.* 2009, 379, (1), 131-8.

[129] Sahoo, S. K.; Diinawaz, F.; Krishnakumar, S., Nanotechnology in ocular drug delivery. *Drug Discovery Today.* 2008, 13, (3-4), 144-151.

[130] Nasr, M.; Mansour, S.; Mortada, N. D.; Elshamy, A. A., Vesicular aceclofenac systems: A comparative study between liposomes and niosomes. *J. Microencapsul.* 2008, 25, (7), 499-512.

[131] Sammour, O. A.; Al Zuhair, H. H.; El Sayed, M. I., Inhibitory effect of liposome-encapsulated piroxicam on inflammation and gastric mucosal damage. *Pharmazeutische Industrie.* 1998, 60, (12), 1084-1087.

[132] Trif, M.; Moisei, M.; Roseanu, A., Designing of efficient lipidic nanostructures for the therapy of the inflammatory diseases. *Romanian Journal of Information Science and Technology.* 2007, 10, (1), 85-95.

[133] Hagiwara, Y.; Arima, H.; Miyamoto, Y.; Hirayama, F.; Uekama, K., Preparation and pharmaceutical evaluation of liposomes entrapping salicylic acid/gamma-cyclodextrin conjugate. *Chemical & Pharmaceutical Bulletin.* 2006, 54, (1), 26-32.

[134] Katare, O. P.; Vyas, S. P.; Dixit, V. K., Enhanced In-Vivo Performance of Liposomal Indomethacin Derived from Effervescent Granule Based Proliposomes. *J. Microencapsul.* 1995, 12, (5), 487-493.

[135] Bula, D.; Ghaly, E. S., Liposome Delivery Systems Containing Ibuprofen. *Drug Dev. Ind. Pharm.* 1995, 21, (14), 1621-1629.

[136] Wang, T.; Deng, Y. J.; Geng, Y. H.; Gao, Z. B.; Zou, H. P.; Wang, Z. Z., Preparation of submicron unilamellar liposomes by freeze-drying double emulsions. *Biochimica et Biophysica Acta-Biomembranes.* 2006, 1758, (2), 222-231.

[137] Maestrelli, F.; Gonzalez-Rodriguez, M. L.; Rabasco, A. M.; Mura, P., Effect of preparation technique on the properties of liposomes encapsulating ketoprofen-cyclodextrin complexes aimed for transdermal delivery. *Int. J. Pharm.* 2006, 312, (1-2), 53-60.

[138] Li, J.; Yang, J.; Wang, W. X.; Yu, J. C.; Fu, J. G.; Wang, X. L., A Novel Liposomal Dexamethasone Palmitate Formulation and Anti-inflammatory Effects on Mice. *Chinese Journal of Chemistry.* 2009, 27, (7), 1411-1414.

[139] Korting, H. C.; Zienicke, H.; Schaferkorting, M.; Braunfalco, O., Liposome Encapsulation Improves Efficacy of Betamethasone Dipropionate in Atopic Eczema But Not in Psoriasis-Vulgaris. *Eur. J. Clin. Pharmacol.* 1990, 39, (4), 349-351.

[140] Piel, G.; Piette, M.; Barillaro, V.; Castagne, D.; Evrard, B.; Delattre, L., Betamethasone-in-cyclodextrin-in-liposome: The effect of cyclodextrins

on encapsulation efficiency and release kinetics. *Int. J. Pharm.* 2006, 312, (1-2), 75-82.
[141] Nii, T.; Ishii, F., Encapsulation efficiency of water-soluble and insoluble drugs in liposomes prepared by the microencapsulation vesicle method. *Int. J. Pharm.* 2005, 298, (1), 198-205.
[142] Martins, S.; Sarmento, B.; Ferreira, D. C.; Souto, E. B., Lipid-based colloidal carriers for peptide and protein delivery--liposomes versus lipid nanoparticles. *Int. J. Nanomedicine.* 2007, 2, (4), 595-607.
[143] Wissing, S. A.; Kayser, O.; Muller, R. H., Solid lipid nanoparticles for parenteral drug delivery. *Adv. Drug Deliv. Rev.* 2004, 56, (9), 1257-1272.
[144] Muller, R. H.; Radtke, M.; Wissing, S. A., Solid lipid nanoparticles (SLN) and nanostructured lipid carriers (NLC) in cosmetic and dermatological preparations. *Adv. Drug Deliv. Rev.* 2002, 54, S131-S155.
[145] Pietkiewicz, J.; Sznitowska, M.; Placzek, M., The expulsion of lipophilic drugs from the cores of solid lipid microspheres in diluted suspensions and in concentrates. *Int. J. Pharm.* 2006, 310, (1-2), 64-71.
[146] Westesen, K.; Bunjes, H.; Koch, M. H. J., Physicochemical characterization of lipid nanoparticles and evaluation of their drug loading capacity and sustained release potential. *J. Control Release.* 1997, 48, (2-3), 223-236.
[147] Bunjes, H.; Westesen, K.; Koch, M. H. J., Crystallization tendency and polymorphic transitions in triglyceride nanoparticles. *Int. J. Pharm.* 1996, 129, (1-2), 159-173.
[148] Muller, R. H.; Radtke, M.; Wissing, S. A., Nanostructured lipid matrices for improved microencapsulation of drugs. *Int. J. Pharm.* 2002, 242, (1-2), 121-128.
[149] Muller, R. H.; Mader, K.; Gohla, S., Solid lipid nanoparticles (SLN) for controlled drug delivery - a review of the state of the art. *Eur. J. Pharm. Biopharm.* 2000, 50, (1), 161-177.
[150] Muller, R. H.; Mehnert, W.; Lucks, J. S.; Schwarz, C.; Zurmuhlen, A.; Weyhers, H.; Freitas, C.; Ruhl, D., Solid Lipid Nanoparticles (Sln) - An Alternative Colloidal Carrier System for Controlled Drug-Delivery. *Eur J Pharm Biopharm* 1995, 41, (1), 62-69.
[151] Souto, E. B.; Müller, R. H., Lipid nanoparticles (solid lipid nanoparticles and nanostructured lipid carriers) for cosmetic, dermal and transdermal applications. In *Nanoparticulate Drug DelivSystems: Recent Trends and*

Emerging Technologies, Thassu, D., Deleers, M., Pathak, Y. (Eds.), CRC Press, Chapter 14, Ed. 2007; pp 213-233.

[152] Souto, E. B.; Müller, R. H., Lipid nanoparticles (SLN and NLC) for drug delivery. In *Nanoparticles for Pharmaceutical Applications*, Domb, A.J., Tabata, Y., Ravi Kumar, M.N.V., Farber, S. (Eds.), American Scientific Publishers, Chapter 5: 2007; pp 103-122.

[153] Cavalli, R.; Bocca, C.; Miglietta, A.; Caputo, O.; Gasco, M. R., Albumin adsorption on stealth and non-stealth solid lipid nanoparticles. *Stp. Pharma Sciences.* 1999, 9, (2), 183-189.

[154] Bhaskar, K.; Anbu, J.; Ravichandiran, V.; Venkateswarlu, V.; Rao, Y. M., Lipid nanoparticles for transdermal delivery of flurbiprofen: formulation, in vitro, ex vivo and in vivo studies. *Lipids in Health and Disease.* 2009, 8.

[155] Han, F.; Li, S.; Yin, R.; Shi, X.; Jia, Q., Investigation of nanostructured lipid carriers for transdermal delivery of flurbiprofen. *Drug Dev. Ind. Pharm.* 2008, 34, (4), 453-458.

[156] Jain, S. K.; Chourasia, M. K.; Masuriha, R.; Soni, V.; Jain, A.; Jain, N. K.; Gupta, Y., Solid lipid nanoparticles bearing flurbiprofen for transdermal delivery. *Drug Deliv.* 2005, 12, (4), 207-215.

[157] Castelli, F.; Puglia, C.; Sarpietro, M. G.; Rizza, L.; Bonina, F., Characterization of indomethacin-loaded lipid nanoparticles by differential scanning calorimetry. *Int. J. Pharm.* 2005, 304, (1-2), 231-238.

[158] Chattopadhyay, P.; Shekunov, B. Y.; Yim, D.; Cipolla, D.; Boyd, B.; Farr, S., Production of solid lipid nanoparticle suspensions using supercritical fluid extraction of emulsions (SFEE) for pulmonary delivery using the AERx system. *Adv. Drug Deliv. Rev.* 2007, 59, (6), 444-453.

[159] Ricci, M.; Puglia, C.; Bonina, F.; Di Giovanni, C.; Giovagnoli, S.; Rossi, C., Evaluation of indomethacin percutaneous absorption from nanostructured lipid carriers (NLC): In vitro and in vivo studies. *J. Pharm. Sci.* 2005, 94, (5), 1149-1159.

[160] Puglia, C.; Blasi, P.; Rizza, L.; Schoubben, A.; Bonina, F.; Rossi, C.; Ricci, M., Lipid nanoparticles for prolonged topical delivery: An in vitro and in vivo investigation. *Int. J. Pharm.* 2008, 357, (1-2), 295-304.

[161] Puglia, C.; Filosa, R.; Peduto, A.; de Caprariis, P.; Rizza, L.; Bonina, F.; Blasi, P., Evaluation of alternative strategies to optimize ketorolac transdermal delivery. *Aaps Pharmscitech.* 2006, 7, (3).

[162] Joshi, M.; Patravale, V., Formulation and evaluation of nanostructured lipid carrier (NLC)-based gel of Valdecoxib. *Drug Dev. Ind. Pharm.* 2006, 32, (8), 911-918.

[163] Attama, A. A.; Weber, C.; Muller-Goymann, C. C., Assessment of drug permeation from lipid nanoparticles formulated with a novel structured lipid matrix through artificial skin construct bio-engineered from HDF and HaCaT cell lines. *J. Drug Deliv. Sci. Technol.* 2008, 18, (3), 181-188.

[164] Xiang, Q. Y.; Wang, M. T.; Chen, F.; Gong, T.; Jian, Y. L.; Zhang, Z. R.; Huang, Y., Lung-targeting delivery of dexamethasone acetate loaded solid lipid nanoparticles. *Arch. Pharm. Res.* 2007, 30, (4), 519-525.

[165] Cavalli, R.; Peira, E.; Caputo, O.; Gasco, M. R., Solid lipid nanoparticles as carriers of hydrocortisone and progesterone complexes with beta-cyclodextrins. *Int. J. Pharm.* 1999, 182, (1), 59-69.

[166] Sivaramakrishnan, R.; Nakamura, C.; Mehnert, W.; Korting, H. C.; Kramer, K. D.; Schafer-Korting, M., Glucocorticoid entrapment into lipid carriers - characterisation by parelectric spectroscopy and influence on dermal uptake. *J. Control Release.* 2004, 97, (3), 493-502.

[167] Liu, W.; Hu, M. L.; Liu, W. S.; Xue, C. B.; Xu, H. B.; Yang, X. L., Investigation of the carbopol gel of solid lipid nanoparticles for the transdermal iontophoretic delivery of triamcinolone acetonide acetate. *Int. J. Pharm.* 2008, 364, (1), 135-141.

[168] Schafer-Korting, M.; Mehnert, W. G.; Korting, H. C., Lipid nanoparticles for improved topical application of drugs for skin diseases. *Adv. Drug Deliv. Rev.* 2007, 59, (6), 427-443.

[169] Paliwal, R.; Rai, S.; Vaidya, B.; Khatri, K.; Goyal, A. K.; Mishra, N.; Mehta, A.; Vyas, S. P., Effect of lipid core material on characteristics of solid lipid nanoparticles designed for oral lymphatic delivery. *Nanomedicine.* 2009, 5, (2), 184-191.

[170] Ruckmani, K.; Sivakumar, M.; Ganeshkumar, P. A., Methotrexate loaded solid lipid nanoparticles (SLN) for effective treatment of carcinoma. *J. Nanosci. Nanotechnol.* 2006, 6, (9-10), 2991-2995.

[171] Cavalli, R.; Gasco, M. R.; Chetoni, P.; Burgalassi, S.; Saettone, M. F., Solid lipid nanoparticles (SLN) as ocular delivery system for tobramycin. *Int. J. Pharm.* 2002, 238, (1-2), 241-5.

[172] Kalam, M. A.; Sultana, Y.; Ali, A.; Aqil, M., Gatifloxacin-loaded solid lipid nanoparticles for topical ocular delivery. *J. Pharm. Pharmacol.* 2009, 61, A75-A75.

[173] Niu, M.; Shi, K.; Sun, Y.; Wang, J.; Cui, F., Preparation of CyA-loaded solid lipid nanoparticles and application on ocular preparations. *J. Drug Deliv. Sci. Technol.* 2008, 18, (4), 293-297.
[174] Attama, A. A.; Reichl, S.; Muller-Goymann, C. C., Sustained Release and Permeation of Timolol from Surface-Modified Solid Lipid Nanoparticles through Bioengineered Human Cornea. *Curr. Eye Res.* 2009, 34, (8), 698-705.
[175] Del Pozo-Rodriguez, A.; Delgado, D.; Solinis, M. A.; Gascon, A. R.; Pedraz, J. L., Solid lipid nanoparticles: formulation factors affecting cell transfection capacity. *Int. J. Pharm.* 2007, 339, (1-2), 261-268.
[176] Friedrich, I.; Reichl, S.; Muller-Goymann, C. C., Drug release and permeation studies of nanosuspensions based on solidified reverse micellar solutions (SRMS). *Int. J. Pharm.* 2005, 305, (1-2), 167-75.
[177] Attama, A. A.; Reichl, S.; Muller-Goymann, C. C., Diclofenac sodium delivery to the eye: in vitro evaluation of novel solid lipid nanoparticle formulation using human cornea construct. *Int. J. Pharm.* 2008, 355, (1-2), 307-13.
[178] Zhang, L. J.; Liu, L.; Qian, Y.; Chen, Y., The effects of cryoprotectants on the freeze-drying of ibuprofen-loaded solid lipid microparticles (SLM). *Eur. J. Pharm. Biopharm.* 2008, 69, (2), 750-759.
[179] Park, E. S.; Chang, S. Y.; Hahn, M.; Chi, S. C., Enhancing effect of polyoxyethylene alkyl ethers on the skin permeation of ibuprofen. *Int. J. Pharm.* 2000, 209, (1-2), 109-119.
[180] Casadei, M. A.; Cerreto, F.; Cesa, S.; Giannuzzo, M.; Feeney, M.; Marianecci, C.; Paolicelli, P., Solid lipid nanoparticles incorporated in dextran hydrogels: A new drug delivery system for oral formulations. *Int. J. Pharm.* 2006, 325, (1-2), 140-146.
[181] Long, C. X.; Zhang, L. J.; Qian, Y., Preparation and crystal modification of ibuprofen-loaded solid lipid microparticles. *Chinese Journal of Chemical Engineering.* 2006, 14, (4), 518-525.
[182] Pang, X. J.; Zhou, J.; Chen, J. J.; Yu, M. H.; Cui, F. D.; Zhou, W. L., Synthesis of ibuprofen loaded magnetic solid lipid nanoparticles. *Ieee Transactions on Magnetics.* 2007, 43, (6), 2415-2417.
[183] Paolicelli, P.; Cerreto, F.; Cesa, S.; Feeney, M.; Corrente, F.; Marianecci, C.; Casadei, M. A., Influence of the formulation components on the properties of the system SLN-dextran hydrogel for the modified release of drugs. *J. Microencapsul.* 2009, 26, (4), 355-364.

[184] Silva, A. C.; Santos, D.; Ferreira, D. C.; Souto, E. B., Oral delivery of drugs by means of solid lipid nanoparticles. *Minerva Biotecnologica.* 2007, 19, (1), 1-5.
[185] Li, X.; Nie, S. F.; Kong, J.; Li, N.; Ju, C. Y.; Pan, W. S., A controlled-release ocular delivery system for ibuprofen based on nanostructured lipid carriers. *Int. J. Pharm.* 2008, 363, (1-2), 177-182.
[186] Barbault-Foucher, S.; Gref, R.; Russo, P.; Guechot, J.; Bochot, A., Design of poly-epsilon-caprolactone nanospheres coated with bioadhesive hyaluronic acid for ocular delivery. *J. Control Release.* 2002, 83, (3), 365-375.
[187] Calvo, P.; RemunanLopez, C.; VilaJato, J. L.; Alonso, M. J., Development of positively charged colloidal drug carriers: Chitosan coated polyester nanocapsules and submicron-emulsions. *Colloid and Polymer Science.* 1997, 275, (1), 46-53.
[188] Shen, J.; Wang, Y.; Ping, Q. N.; Xiao, Y. Y.; Huang, X., Mucoadhesive effect of thiolated PEG stearate and its modified NLC for ocular drug delivery. *J. Control Release.* 2009, 137, (3-4), 217-223.
[189] Gasco, M. R.; Gallarate, M.; Trotta, M.; Bauchiero, L.; Gremmo, E.; Chiappero, O., Microemulsions as topical delivery vehicles: ocular administration of timolol. *J. Pharm. Biomed. Anal.* 1989, 7, (4), 433-9.
[190] Wong, H. L.; Li, Y.-Q.; Bendayan, R.; Rauth, M. A.; Wu, X. Y., Solid Lipid Nanoparticles for Antitumor Drug Delivery. In *Nanotechnology for Cancer Therapy*, Mansoor M. Amiji (Ed.), CRC Press, Chapter 36, 741-776: 2007.
[191] Müller-Goymann, C. C., Physicochemical characterization of colloidal drug delivery systems such as reverse micelles, vesicles, liquid crystals and nanoparticles for topical administration. *Eur. J. Pharm. Biopharm.* 2004, 58, (2), 343-356.
[192] Ludwig, A., Ocular Applications of Nanoparticulate Drug-Delivery Systems. In *Nanoparticulate Drug DelivSystems: Recent Trends and Emerging Technologies*, Thassu, D., Deleers, M., Pathak, Y., Ed. CRC Press: 2007; pp 271-280.
[193] Date, A. A.; Joshi, M. D.; Patravale, V. B., Parasitic diseases: Liposomes and polymeric nanoparticles versus lipid nanoparticles. *Adv. Drug Deliv. Rev.* 2007, 59, (6), 505-521.
[194] Siqueira, R. C.; Filho, E. R.; Fialho, S. L.; Lucena, L. R.; Filho, A. M.; Haddad, A.; Jorge, R.; Scott, I. U.; Cunha Ada, S., Pharmacokinetic and toxicity investigations of a new intraocular lens with a dexamethasone

drug delivery system: a pilot study. *Ophthalmologica.* 2006, 220, (5), 338-42.

[195] Farshi, F. S.; Ozer, A. Y.; Ercan, M. T.; Hincal, A. A., In-vivo studies in the treatment of oral ulcers with liposomal dexamethasone sodium phosphate. *J. Microencapsul.* 1996, 13, (5), 537-44.

Index

A

acetic acid, 10, 36
acid, 8, 10, 17, 21, 22, 24, 25, 28, 44, 49
adhesion, 10, 38
alcohol, 12, 20, 25
alcohols, 9, 25
allergic conjunctivitis, 3, 4, 10, 34
allergic reaction, 8
amines, 8
analgesic, 8, 10
anti-inflammatory agents, 3, 4, 8, 9, 19, 21, 31, 37
anti-inflammatory drugs, vii, 5, 11, 15, 19, 21, 23, 27, 31, 34
antiviral agents, 18
aqueous humor, 13
aqueous solutions, 12, 17
aqueous suspension, 20
authors, 14, 15, 16, 17, 19, 20, 23
autoimmunity, 3, 7
availability, 38
avoidance, 10, 21, 26

B

barriers, vii, 4, 10, 35

bioavailability, vii, 5, 11, 13, 14, 15, 19, 20, 23, 24, 27, 28, 40
biocompatibility, 6
biodegradation, 25
blood, 3, 4, 7, 8, 10, 18
blood flow, 3
brain, 19
burn, 16, 40
burning, 11, 26

C

capillary, 9
carcinoma, 47
carrier, 5, 15, 47
cataract, 4, 7, 9, 10, 33, 34, 37, 38
cataract extraction, 7, 37
cell, 8, 16, 18, 23, 47, 48
cell line, 23, 47
cell lines, 47
cell membranes, 8, 16, 18
cholesterol, 17, 18, 19
choroid, 8
choroiditis, 7
colonization, 10, 38
combination therapy, 37
compliance, 11, 13
complications, 4, 7
composition, 15, 18, 21, 23, 24, 41
compounds, 3, 5, 9, 10, 18, 19, 20, 21, 23

concentration, 5, 8, 12, 13, 18, 19, 20, 23, 24
conjunctiva, vii, 7, 8, 16, 20
conjunctivitis, 3, 33
contact time, 11
control, 4, 9, 18, 27, 38
cornea, vii, 5, 7, 8, 9, 10, 12, 16, 17, 20, 23, 24, 39, 48
corticosteroids, vii, 3, 5, 13, 19, 34, 36, 37
costs, 12
crystals, 22
cyclodextrins, 44, 47
cyclooxygenase, 8
cyclosporine, 23, 25, 41
cystoid macular edema, 34, 37, 38
cytokines, 8, 9
cytomegalovirus, 42
cytomegalovirus retinitis, 42

D

defects, 4, 10
degradation, vii, 11, 18, 25, 27
delivery, vii, 5, 11, 12, 14, 15, 16, 18, 19, 20, 23, 24, 26, 27, 28, 29, 31, 33, 35, 36, 39, 41, 42, 43, 44, 45, 46, 47, 48, 49
differential scanning, 46
differential scanning calorimetry, 46
discomfort, 4, 7, 10, 11, 13, 38
dispersion, 21, 25
distribution, 18, 19, 35, 40, 43
DNA, 9, 23, 42
dosage, 5, 14, 16
drainage, 5, 11
drug carriers, 5, 40, 49
drug delivery, v, vii, 5, 6, 11, 12, 13, 20, 21, 24, 25, 27, 35, 37, 38, 39, 40, 41, 43, 44, 45, 46, 48, 49, 50
drug release, 4, 15, 18, 20, 22, 24, 26, 27, 31
drug therapy, 4
drug use, 23
drugs, vii, 3, 5, 8, 9, 10, 11, 12, 18, 19, 22, 23, 27, 28, 29, 31, 37, 38, 39, 45, 47, 48, 49
drying, 44, 48

E

edema, 35
electron, 41
emulsions, 11, 13, 14, 16, 23, 39, 40, 44, 46, 49
encapsulation, 18, 20, 21, 45
enzymes, 5, 8, 10
epithelial cells, 24, 38
epithelium, 5, 9, 15, 16, 20
equilibrium, 14
ester, 20
ethers, 48
evaporation, 20
expulsion, 21, 45

F

fatty acids, 21, 25
FDA, 1, 9, 10
fibrin, 7
fibrosis, 7
first generation, 21
fluid, vii, 5, 10, 12, 13, 14, 15, 16, 40, 46
fluid extract, 46
fluorescence, 14

G

gastric mucosa, 44
gel, 33, 47
gene, 23, 42, 43
gene therapy, 23
gland, 37
glaucoma, 3, 9, 43
glucose, 28
glycol, 25
goblet cells, 16
guidelines, 37

H

half-life, 24

healing, 10, 17, 40
health, 3
herpes simplex, 10
hydrocortisone, 23, 47
hydrogels, 36, 38, 48
hydrolysis, 10
hyperemia, 10
hypersensitivity, 3, 7, 10
hypertension, 9
hypothesis, 14, 16

I

ibuprofen, 21, 24, 48, 49
ideal, 24, 31
immune response, 9
immunotherapy, 33
implementation, 36
in vitro, 15, 16, 20, 38, 40, 41, 43, 46, 48
in vivo, 5, 15, 25, 46
infection, 4, 10, 31
inflammation, 3, 4, 7, 8, 9, 10, 33, 37, 38, 44
inflammatory disease, v, 7, 11, 12, 21, 23, 31, 33, 44
inflammatory mediators, 3, 7, 8, 9
inhibition, 9, 10, 17, 37
injections, 4, 12, 18
injuries, 34
injury, iv, 3, 7, 8, 36, 37
integrity, 37
interaction, 11, 16, 18, 24
interactions, 15
interface, 12, 14
interfacial layer, 12
intraocular, 4, 5, 7, 10, 37, 49
intraocular pressure, 7, 10
ionization, 15
iris, 7, 8, 20, 37
iritis, 7
ischemia, 3

K

keratoconjunctivitis, 34
keratoplasty, 35
kidney, 19
kinetics, 45

L

leakage, 7, 22
lecithin, 15, 17, 18, 28
lens, 7, 8, 49
leukotrienes, 8
limitation, 15
lipids, 6, 18, 21, 22, 25
liposomes, vii, 6, 12, 17, 18, 19, 20, 21, 23, 41, 43, 44, 45
liquid crystals, 49
liver, 19
localization, 21
lysozyme, 20

M

macrophages, 9
maintenance, 8, 18
management, 4, 10, 33, 34
matrix, 22, 25, 47
media, 18, 38
melting, 10, 22
membranes, 17, 18, 35
metabolism, 5
metabolites, 8
mice, 19, 43
microemulsion, 13, 39
microscope, 41
microspheres, 45
miosis, 4, 7, 8, 9, 10, 34
model, 22, 37, 41
models, 22
molecular weight, 20, 25, 43
molecules, 9, 10, 18, 22
mucin, 15
mucosa, 24

mucus, 25

N

nanometric range, 14
nanoparticles, vii, 6, 15, 16, 21, 22, 23, 25, 29, 36, 40, 45, 46, 47, 48, 49
nanostructures, 44
nerve fibers, 36
neuropeptides, 8
neutrophils, 9
non-steroidal anti-inflammatory drugs, vii, 3, 34
NSAIDs, 1, 3, 4, 5, 8, 10, 33
nucleus, 9

O

ocular diseases, 4, 23, 31
oedema, 4, 9, 10
ofloxacin, 42
oil, 12, 13, 14, 15, 16, 22, 38
optic nerve, 10
optimization, 40, 41
order, 15, 18, 23, 24, 25
organic solvents, 21

P

pain, 3, 4, 7, 8, 10, 36, 38
parameters, vii, 13
particles, vii, 21, 23, 24
partition, 5, 15, 23
pathogens, 3
pathways, 9
patient care, 37
permeability, 5, 8, 9, 10, 14, 17, 18, 24
permeation, 5, 10, 15, 23, 39, 47, 48
pharmacokinetics, 38, 41, 42, 43
pharmacological treatment, 31
pharmacology, 37
phosphatidylcholine, 19, 28
phospholipids, 8, 15, 17, 18, 20, 24, 25, 41
physicochemical properties, 5, 17

pigmentation, 7
pilot study, 50
placebo, 37
plasma, 18
plasmid, 42
polymer, 24
polymers, vii
poor, 5, 11, 24, 26
prevention, 4, 10, 34
prodrugs, 5
production, 6, 8, 9, 10, 11, 12, 18, 21, 26, 27, 28, 31, 40
progesterone, 47
proliferation, 9
prophylaxis, 9
prostaglandins, 7, 8, 10
proteins, 9
ptosis, 9

R

range, 15, 25
refractive index, 13
retention, 11, 12, 20, 24, 25, 26, 39
retina, vii, 7, 8
retinal detachment, 4
retinal disease, 23
retinitis, 7
risk, 4, 9, 10

S

safety, 27, 34
sclera, vii, 8, 10, 20
secretion, 5
sensation, 4, 11, 26, 34
shape, 17, 19
side effects, vii, 4, 5, 9, 10, 18, 24, 31
skin, 47, 48
skin diseases, 47
smooth muscle, 8
sodium, 15, 20, 23, 24, 25, 27, 34, 38, 43, 48, 50
soft matter, 39

solid matrix, 23
solubility, 5, 24, 27
spectroscopy, 47
stability, 12, 14, 18, 20, 21, 26, 27, 29, 31, 39
sterile, 39
steroids, 9, 10, 20, 35, 37
stoma, 10
storage, 12, 21, 22
strategies, vii, 46
stroma, 9
sugar, 19, 43
surface modification, 23, 24
surface properties, vii
surface tension, 12
surfactant, 13, 15, 21, 25, 27
suspensions, 21, 45, 46
swelling, 3
symptoms, 3
syndrome, 42
synthesis, 8

T

therapeutics, 29
therapy, 5, 33, 34, 36, 37, 42, 44
thromboxanes, 8
tissue, 3, 7, 11, 35, 40

toxicity, 4, 6, 13, 14, 18, 20, 25, 26, 27, 31, 42, 49
toxicology, 42
transfection, 23, 42, 48
transitions, 22, 45
transport, vii, 15
triglycerides, 22

U

ulcer, 17
uveitis, 3, 7, 33, 34, 36, 37

V

vehicles, 11, 43, 49
vesicle, 18, 21, 45
vessels, 7
viscosity, vii, 5, 13, 25
vision, 3, 4, 11, 13, 14, 26
visual acuity, 7, 10
visual field, 10

W

wettability, 16
wound healing, 10